LIFE SCIENCE LIBRARY

GIANT
MOLECULES

TIME
LIFE
BOOKS
®

LIFE WORLD LIBRARY

LIFE NATURE LIBRARY

TIME READING PROGRAM

THE LIFE HISTORY OF THE UNITED STATES

LIFE SCIENCE LIBRARY

GREAT AGES OF MAN

TIME-LIFE LIBRARY OF ART

TIME-LIFE LIBRARY OF AMERICA

FOODS OF THE WORLD

THIS FABULOUS CENTURY

LIFE LIBRARY OF PHOTOGRAPHY

THE TIME-LIFE ENCYCLOPEDIA OF GARDENING

THE AMERICAN WILDERNESS

THE EMERGENCE OF MAN

THE OLD WEST

FAMILY LIBRARY:
   THE TIME-LIFE BOOK OF FAMILY FINANCE
   THE TIME-LIFE FAMILY LEGAL GUIDE

LIFE SCIENCE LIBRARY

CONSULTING EDITORS
René Dubos
Henry Margenau
C. P. Snow

GIANT MOLECULES

by Herman F. Mark

and the Editors of **TIME-LIFE BOOKS**

TIME-LIFE BOOKS   NEW YORK

ABOUT THIS BOOK

SYNTHETIC GIANT MOLECULES have revolutionized the technology of the world. The ability of chemists to design giant molecules, or polymers, to meet specific needs, and then to produce them according to plan, has led to an avalanche of new products with qualities that could never be duplicated with natural materials. Almost every aspect of contemporary life —from space flight to oceanography, from surgery to agriculture, from electronic computers to suburban homes—reflects this scientific breakthrough which began scarcely a half century ago. This book tells the story of man-made giant molecules, of the centuries of progress in chemistry that led to their creation, and of the spectacularly swift growth of research and mass production during the past several decades.

Each text chapter has a supplementary picture essay that may be read independently. For example, Chapter 7, "Molecules Made to Order," deals with chemists' ability to make polymers with almost any desired characteristics; it precedes a picture essay, "Plastics' New Bag of Tricks," which shows the spectacular properties of many of these synthetics.

THE AUTHOR

HERMAN F. MARK is one of the greatest of the pioneers of polymer chemistry. As head of the eminent research team assembled by Germany's I. G. Farben Company in the 1920s, he supervised work that threw new light on the nature of the giant molecule, and led to the creation of a host of commercially valuable polymers, such as polystyrene. Before World War II, Dr. Mark left Germany and became a professor at the Polytechnic Institute of Brooklyn, where he organized the world-famous Polymer Research Institute and founded the *Journal of Polymer Science*. He is the author of 14 books and more than 400 articles on chemistry. He holds 10 honorary degrees, is a member of many academies and scientific societies and is now Dean of the Faculty Emeritus at the Polytechnic Institute of Brooklyn.

THE CONSULTING EDITORS

RENÉ DUBOS, a member and Professor Emeritus of The Rockefeller University, is a distinguished microbiologist and experimental pathologist who won the Arches of Science Award in 1966 and the Pulitzer Prize in 1969 for his book *So Human an Animal: How We Are Shaped by Surroundings and Events*. He also authored *Mirage of Health* and *Man Adapting* and co-authored *Health and Disease* in this series.

HENRY MARGENAU is Eugene Higgins Professor Emeritus of Physics and Natural Philosophy at Yale and an authority in the fields of spectroscopy and nuclear physics. He wrote *Open Vistas, The Nature of Physical Reality*, and is co-author of *The Scientist* in this series.

C. P. SNOW has gained an international audience for his novels, including *The New Men, The Affair* and *Corridors of Power*, which explore the influence of science on today's society.

ON THE COVER

The giant molecule of Lexan plastic displays its hexagonal structure, deliberately created by chemists to give the substance its great strength. Shown on the back cover is a stylized depiction of small molecules being converted into giant molecules in the laboratory.

Library of Congress catalogue card number 66-19119.
School and library distribution by Silver Burdett Company, Morristown, New Jersey.

CONTENTS

INTRODUCTION

IF OUR AGE were to be named for the materials that characterize it—as were the Stone and Bronze Ages of the past—this might be known as the age of plastics. For plastics, made of synthetic giant molecules, have become a dominating influence on modern industrial society.

Giant molecules, or polymers, are the stuff of life. They constitute the material that cells are made of, the protoplasm and nuclei of both animal and plant tissue. Nevertheless, their structure—and the effect of this structure on their properties—was unknown until only a few decades ago.

It was not until the early 1920s that polymer chemistry became a separate branch of chemistry in general. Thanks to the research started at that time in universities (by Hermann Staudinger) and in industrial laboratories (by, among others, Herman Mark, the author of this book), it was revealed for the first time that polymers existed in distinctive long, orderly chains linking smaller molecules. With this knowledge researchers could begin to create polymers of their own, including many new ones that had never before existed in nature. These discoveries were the basis for the now enormously important plastics, synthetic fiber and synthetic rubber industries. Although the work began initially in Europe, the great U.S. chemical firms were among the first to get into production with certain synthetic polymers such as nylon and the synthetic rubber, neoprene. Polymer chemistry became literally vital to the United States and its Allies during World War II when the Japanese occupation of Malaysia and the Malay Archipelago cut off almost all supplies of natural rubber. Only the ability of the U.S. to produce synthetic rubber in huge quantities saved the Allied military cause.

The synthetic production of giant molecules is bound to expand in the coming years, and as it does so it will also affect critical problems of society. For example, increased production of synthetic textiles and rubber could free vast tracts of land for the growing of food—land now used to grow natural rubber and textile fibers (cotton and wool). It will thus be possible to help relieve the food scarcities which threaten to grow worse as world population increases.

This book, *Giant Molecules*, provides an excellent, clearly written description of the developments in chemistry which led first to an understanding of molecules in general, then to giant molecules, and finally to the mass production of man-made polymers—and the plastics age in which we live.

—DR. GIULIO NATTA
Professor of Industrial Chemistry
Polytechnic Institute, Milan
Nobel Prize winner in Chemistry, 1963

1

An Age
of
Plastics

Ribbons of man-made acrylic fiber, used for carpets and clothing, roll through a factory.

AS THE WELL-DRESSED MATRON rummaged frantically for her commuter ticket in her overfilled handbag, the conductor stared at the ceiling with the patience of a man who has already suffered much and is convinced that the future will be no brighter. Years of practice had given him mastery of this performance with passengers who did not have tickets ready, and he was gloomily pleased with the anger he could feel radiating from this woman. Finally she found the ticket and thrust it at him as though it were a dagger. As he walked away, she vigorously snapped her purse closed, not noticing the comb and pencil which projected above the opening. The plastic clasp broke. She stared at the broken fastener for a moment, and then said in a voice that carried throughout the club car, "I . . . hate . . . all . . . plastics!"

The tall, dignified man seated on her right lowered his newspaper. "You do not hate all plastics," he said firmly. "You would find contemporary life insupportable without plastics."

"On the contrary, I would find life delightful," she said. "Plastics are synthetics—cheap imitations, always breaking or cracking at the wrong time. People got along beautifully without synthetics for centuries."

"Perhaps not so beautifully," the man murmured. "I doubt that you would care to return to that earlier day. As a matter of fact, if all plastics were to disappear at this moment, you'd find yourself in a rather awkward predicament. From the lifts on your heels to the spray that maintains your attractive hair arrangement, you are a living advertisement for synthetics. Your stockings are undoubtedly nylon, your handsome suit is Orlon, and the other—ah—garments you are wearing are, in all probability, synthetics—spandex, most likely."

"Upon consideration," she said thoughtfully, "I can see that I might find things a bit drafty."

"Your home is equally a part of the plastic age," he said. "The tops of your kitchen table and counters are, almost certainly, sheathed in plastic, you may cook with fluorocarbon-lined pans, and your polystyrene-lined refrigerator is probably filled with foods that are kept fresh with polyethylene wrapping.

"The other areas of your home are also indebted to plastics or other synthetics. Your stain-resistant carpet, for example, and the cabinet of your television set, perhaps the covering on some of your furniture—all synthetics. A modern bathroom may have walls and floor of plastics, and certainly contains dozens of other synthetic items, from the shower curtain to toothbrushes. Even your automobile has literally hundreds of plastic parts. The reason modern cars require fewer greasings is that they have nearly friction-free plastic bearings."

"Speaking unkindly of plastics when you're around," the woman said with a note of awe, "is like kicking a lion cub in the presence of its mother. Do you, by any chance, own a plastics company?"

"I do not," the man said with hauteur. "I am a chemist, a polymer chemist, which means I work with giant molecules. All plastics are composed of giant molecules."

"I didn't know they came in two sizes," the woman said. "What's the difference between the giants and small ones?"

"The small molecule," the chemist said, "is the molecule which comprises most of the inanimate things of the world—water, minerals, air. Giant molecules, or polymers as they are called, are made up of repeating units of small molecules, and occur only in living matter—plants and animals, including man—and in certain synthetics. Their outstanding properties are great molecular weight and a chainlike structure."

The woman cleared her throat self-consciously. "At the risk of shocking you," she said, "I must ask a question. Exactly what is a molecule?"

"Perhaps I should begin a step or two before that," the chemist said. "Polymer chemistry is a fascinating story. Let me tell you something about it during the remainder of our ride. Let's begin with a basic subject: the elements."

All matter on earth, the chemist explained—all solids, liquids and gases—is made up, as far as is known, of one or more of some 100 elements. Astronomers have detected a number of these elements on other planets, on stars and in meteors which have fallen to earth from outer space. Thus far, there is no indication that the celestial bodies contain any elements unknown to earth. This has led scientists to the belief that the same basic materials comprise the entire universe.

Visualizing the unseeable

An element is a "pure" thing; it cannot be broken down any further by chemical means. It is, in a sense, like a primary color that is composed of nothing but itself. The smallest part of an element is the atom. No one has ever seen an atom—it is far too small—but brilliant inductive reasoning by a number of men plus X-rays and mathematical tools have given science a good idea of the atom's structure: a dense core with clouds of electrons whirling constantly about it.

However, atoms are gregarious. They seldom exist in a solitary state, but through electrical force they band together, sometimes with their own kind, sometimes with other elements. An atom of oxygen, for example, may join up with another atom of oxygen to form a molecule of oxygen. But if the oxygen atom combines with a different element, a molecule that is a compound is created. Water is such a compound. One atom of oxygen (O) has joined up with two atoms of hydrogen (H); in the chemist's shorthand, it is H_2O. Like an element, a compound is a pure thing; every part has exactly the same composition down to the smallest possible division.

One fascinating aspect of compounds is that each of them has properties not shared by the elements that make it up. Water, for example, is composed of two gases. One of them, hydrogen, is flammable; the other, oxygen, is necessary to combustion. Yet water is a liquid; it will not burn nor will it support combustion. There are literally millions of compounds that make up most of the matter on earth. However, about 99

AN ATOMIC THEORIST who lived in the Fifth Century B.C., the Greek Democritus was one of the first men to express the theory that matter is made of atoms. Although he used only simple observation and logic as the basis for his reasoning, several of his theories—for example, that atoms are identical in substance and differ in size and weight—have been verified by modern scientific investigation.

per cent of matter having to do with living things is made up of a mere handful of elements—hydrogen, oxygen, nitrogen, carbon and sulfur—which comprise some million and a half compounds.

But the most common of all the elements in compounds is carbon. It occurs with such frequency, in fact, that chemistry is usually divided into two major fields: organic, involving compounds that contain carbon, and inorganic, involving compounds that do not. The carbon atom is one of the few elements with four bonding points, which connect it with other atoms. Most other elements have fewer. Carbon joins up with a wide variety of elements, and forms long chains, the basic structure of polymers. Thus, carbon is found in millions of compounds, including all substances that are a part of, or come from, living matter.

Independence within mixtures

Mixtures, like compounds, are also combinations of two or more kinds of atoms. However, in a mixture, the atoms of the ingredients are not bonded together with an electrical force as they are in a compound. Each component retains its identity and characteristics, and the parts can readily be separated by mechanical means. The food man eats, many of the clothes he wears, the air he breathes, his very blood and bones—all of these are mixtures. None of the constituents are atomically joined to form a new substance. Air, for example, is made up principally of nitrogen and oxygen, but each gas retains its independent characteristics and, with proper equipment, can be separated from the mixture by purely physical processes.

Mixtures are usually identified as either heterogeneous or homogeneous. The latter word is often used to describe one or more substances in solution. In a heterogeneous mixture, at least some of the components can often be identified. In a slab of concrete, for example, bits of rock, sand and cement are recognizable. In contrast, a homogeneous mixture such as a solution of sugar and water looks like ordinary water. However, if the water evaporates, sugar crystals will remain in the container.

The rigorous classification of matter into elements, compounds and mixtures was achieved only after hundreds of years of study and experiment, and not until the 20th Century was it firmly established that there are two distinct families of molecules—small and giant molecules, or polymers. The small molecule, which comprises all inanimate things—air, water, metal, stone—is light in molecular weight. The heavier giant molecule is made up of small molecules, sometimes numbered in the hundreds of thousands, and linked together in a repeating pattern.

Natural giant molecules are the stuff of life. Everything that lives and grows—both animal and vegetable—is made up largely of polymers. Nature has been creating them since the beginning of life on this planet, but only within this century has man been able to duplicate nature's feat of making giant molecules from small ones. And with this great discovery has come an entire synthetics industry that plays an enormous role in contemporary life.

Men have been using giant molecules, of course, during all the years of the human race. The proteins of meat are giant molecules—among the largest of all, in fact. In the process of eating, the human digestive system disassembles this molecule, only to reconstitute it into other giant molecules essential to life.

In the earliest times man could do nothing with the polymers of nature but use them more or less as he found them. He ate the uncooked flesh of the animals he killed, and then used their raw skins for clothing. He supplemented his diet with the fresh fruit and berries he found around him. But then he learned to improve on nature by manipulating her materials mechanically. He could weave fibers like silk and wool into cloth, and cut wood into pieces to build effective shelter. The first man to build a fire was initiating a chemical change in cellulose, the ubiquitous giant molecule of the plant world. By burning wood he enabled the oxygen of the air to break down the cellulose in the wood and at the same time release various gases, carbon, light and heat. He knew nothing about all that, of course; he knew only that fire warmed him.

Progress through fire

The discovery of controllable fire enabled men to go further in the use of natural polymers. Now it became possible to alter their characteristics in a purposeful way, through chemical action. The first such deliberate chemistry was cooking, an invention whose significance often goes unrecognized. Meat tastes better and is more easily digested because heat sets off chemical reactions that alter and degrade its molecules. This interference with a polymer of nature not only adds to the pleasures of human existence, but prolongs life. Cooked food sustains the very young and the very old, the weak and the sick, who might not survive if they had to eat raw meat.

Though the chemical process of cooking was long ago developed into a fine art, it never became a science: even today, the exact reactions involved are not completely understood. Nevertheless, for thousands of years it was one of the few ways that the raw polymers of nature could be improved upon chemically.

The techniques of ancient chemistry were carried out with praiseworthy skill. But it was only skill. There was no theory to suggest new approaches or to explain the frequent failures. An improvement was an accident, often impossible to duplicate. Not until science replaced skill —and in particular, until science was based solidly on atomic theory— could men do more with polymers than use them much as they were found in nature.

The essential idea of the atom was conceived and then put aside more than 2,000 years ago. About 400 B.C. a Greek philosopher named Democritus postulated that matter was not continuous but was made up of a number of tiny, constantly moving particles; he called them "atoms," from the Greek word *atomos*, meaning "indivisible."

Democritus' theory was not as inspired a flight of fancy as it might seem. It was a commonsense deduction based on ordinary, observable phenomena. Salt or sugar disappears when dissolved in water. If water were continuous, if it were not made up of small particles, there would be no space into which the salt or sugar could disappear. Similarly, if a dye is put into clear water and the mixture is stirred, an evenly colored solution results. Obviously the commingling of the two substances can be explained only if the water and dye are composed of minute units, with space between them for the mixing to take place. Perhaps the Greek philosopher even poured a gallon of alcohol into a gallon of water and found that the combination totaled less than two gallons. The reason for this paradox, scientists now know, is that the alcohol occupies spaces between the particles of water, and the water moves into spaces between the alcohol particles, causing the combination to shrink. Such a man as Democritus could have deduced the reason. Unfortunately, a number of his more famous contemporaries dismissed Democritus' atomic theory and advanced some wrongheaded concepts of their own, which were accepted as absolute truth for many centuries.

This acceptance of gross error as dogma was at least partially responsible for the failure of chemistry to make any substantial advances until about the 18th Century. Only then—as man began to throw off the shackles of misinformation, to make controlled experiments and to learn some of the natural laws he had been applying for generations—did chemistry evolve from an ancient art to a true science.

The mystery of living substance

An enormous amount of knowledge was gained during the years that followed, but most of it was in the field of nonliving matter—the small molecule. Living substances proved a far more baffling subject. The problem was that no one understood the basic structure of the giant molecule, or how it differed from the small molecule. By the 1800s, chemists could separate and reassemble many of the ordinary molecules that are present in soil, water, minerals and the gases of the atmosphere, but attempts to analyze materials that came from living substances were unsuccessful. The tests ordinarily used with nonliving materials simply did not work. When, for instance, a substance from life was changed to something else by heat, there was no way of restoring it to its original form. Whereas a chemist could decompose water into hydrogen and oxygen and then recombine the gases to create water, he could do nothing about converting wood ashes back to a log. Because scientists could learn little about the giant molecule—in fact, they did not even know of its existence—they could do nothing about changing the materials from life. They could only follow the centuries-old practice of improving natural products. A crude raincoat could be made by treating cloth with oil, for example, and the life of lumber could be prolonged with tar or formaldehyde, but no change could be wrought in the basic materials themselves.

Not until the 19th Century, when Democritus' old theory of atoms

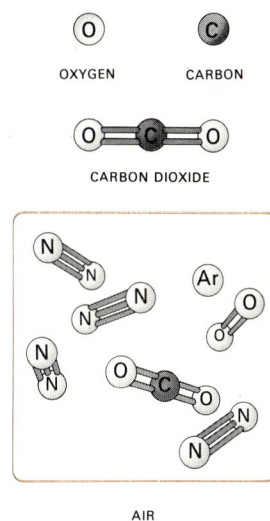

THREE FORMS OF MATTER—elements, compounds and mixtures—are illustrated here. The simple elements carbon and oxygen are shown as single atoms *(top)*, and also bonded together in the compound carbon dioxide *(center)*. A mixture, such as air *(bottom)*, occurs when compounds or elements come together without losing their chemical identities.

13

could be applied to living matter, was real progress made. There followed the rapid, even riotous, succession of revelation and invention that continues today. Brilliant deductions, occasionally mystical in their inspiration, alternated almost at random with accidental discoveries. A lucky chemist would stumble on a valuable product or process, sometimes in a purposeful search, sometimes by mistake. The results of his experiment would stimulate a theoretician to a new and deeper elucidation of the nature of molecules. Testing the theories led other experimenters to still more discoveries. A student, hoping to duplicate a natural drug, instead synthesized from coal tar a brilliant dye of a color that never existed in nature. Another young chemist visualized in a dream the basic layout of atoms that could be applied to polymers.

A profit-making accident

Acid spilled in a kitchen led to the first alteration of the giant molecule of cellulose, making possible a host of new products: guncotton for explosives; collodion, a dressing for wounds; cellophane, still one of the most useful packaging materials; celluloid, which was used for everything from knife handles to gentlemen's collars; the fibers viscose rayon and acetate rayon, which could be woven into lustrous, imitation silk.

Gradually the rational laws of chemistry that had been codified over centuries of work on simple small molecules were extended to apply to larger and more complex ones. But none of the new materials developed thus far was a true synthetic. The basic properties of all the cellulose-based substances were derived in altered form from a previously existing polymer. The task of creating a completely new polymer from small molecules, of building a giant molecule unlike any in nature, remained to be done.

When this goal was finally achieved—by the Belgian-American Leo Baekeland, who developed the first synthetic plastic, Bakelite, in 1907 —an entirely new scientific world opened up. The rapidity with which this virgin territory was explored was breathtaking. Theory began to catch up with practice, thus hastening the process. Intensive polymer research began in Europe and the United States. Soon chemists learned to characterize giant molecules by their length-per-chain and to analyze their structures. Researchers found that polymers differ from small molecules chiefly in weight and design. In the past, many chemists had theorized that a mysterious "life force" provided the connecting links within the giant molecule. Now they knew that the giant molecule is held together by the same forces that bond ordinary substances.

Then industry entered the picture, first in Germany, later in the United States and other countries. Seeing the enormous possibilities of polymer synthetics, manufacturers recruited theoretical polymer chemists from universities and private laboratories and provided these men with all the equipment they needed for research. Soon chemists understood the polymer so well that they could plan in advance the giant molecules that could be made from small molecules.

This was a turning point in the history of science. Until the 1920s, chemists had been practicing a kind of "cookbook chemistry" in dealing with polymers. Workable recipes were found on a trial-and-error basis and then passed on. Another chemist-cook might add to or subtract from the recipe, but his work, too, was almost entirely empirical. After learning about the structure of the giant molecule and how it could be put together, the chemist became a kind of architect; he could, in many instances, design the molecule to incorporate the properties he wanted it to have before he ever began to create it.

Out of this understanding of polymers came a gigantic, and entirely new, chemical industry. Polymer chemistry affected almost every manufacturing business and touched the lives of people in every part of the world. The ability of engineers to translate the findings of the laboratory into mass production at low cost was, of course, vital to this new industrial revolution. As a matter of fact, the skill of American engineers in making great quantities of synthetic rubbers—all of them polymers—was literally decisive in the outcome of World War II.

Chemistry's global revolution

Since then, the effects of polymer chemistry and the mass production of synthetics have changed the world. Everything from surgery to space travel utilizes plastics. Man's home, his clothing, the cars, trains and planes that carry him, a thousand things he touches and uses every day are, to a great extent, the result of polymer chemistry.

And just over the horizon are inexpensive houses molded in a matter of hours, plastic parts of the body to replace those which are diseased, collectors of solar energy to supply mankind's ever-increasing need for power, whole libraries reduced to film that will fit into a filing cabinet, perhaps entire cities enclosed in giant plastic envelopes.

The train was pulling into the station as the chemist finished his lecture, and he began to stuff his newspaper into his briefcase. "Thank you for an educational trip," the woman said. Then she glanced down at the purse bulging open on her lap. "Perhaps what I said about hating all plastics was a bit strong, but I still think there are a lot of them of poor quality."

"Of course there are," the chemist said. "There are cheap, brittle plastics just as there are inferior goods of all kinds. But if the manufacturer of that bag had wanted to do his job properly, he could have used a plastic that would have outlasted almost any metal."

The chemist reached into his briefcase and pulled forth a large rubber band. "Slip this around your purse," he said. "It should hold it closed until you can get a new clasp."

"Thank you," the woman said.

The chemist rose and put on his hat. "The band is synthetic rubber, of course," he said smugly. Then he nodded pleasantly and walked out of the car.

Wigs, Dams
and Statues

This is an age of synthetics. Today a stylish woman can go to the theater completely attired in synthetic material, from her acrylic wig *(right)* to her vinyl soles. Tomorrow a businessman may be able to leave his all-plastic house in the morning, walk across a lawn of polyethylene grass to his fiber glass car, and drive to work over roads cushioned with synthetic rubber. After centuries of relying on nature for the same materials for shelter, clothing and household goods, man has turned to chemistry for substitutes that are often superior to their natural prototypes. The huge and growing American plastics industry molds, extrudes, draws and laminates some 50 different types of synthetic compounds, or polymers, and sells more than $5.5 billion worth of them annually. Not only can chemists manipulate molecules to reproduce virtually any substance derived from a living organism; they can go a step further and create entirely new materials unlike any seen before. Synthetics, their versatility barely tapped by the thousands of products now in use—including folding dams and marblelike statues—face a future as limitless as the ingenuity of the men who create them.

A WATERPROOF WIG

Dynel hair, as soft and lustrous as the human variety, has the added advantage of being so moisture-resistant that it will hold a hairdo in a rainstorm. Like other members of the acrylic family of synthetic fibers, which are widely used in clothing fabrics, Dynel is made from natural gas, salt, ammonia and water, and can simulate textures from fluffy cashmere to a bristly mohair.

A SYNTHETIC SETTING
In the dining area of the model plastic home below, the only nonplastic material in sight is the safety glass of the windows and divider—and that is bonded by thin strips of plastic. Furniture and decorating materials were selected from a great variety of plastic products.

A MOLDED RESIDENCE
The world's first plastic house, built by M.I.T. and Monsanto in 1957, had six rooms and consisted of 16 molded segments, each containing a half-ton of polyester resin. After 10 years and over 20 million visitors, the unsupported wings sagged less than a quarter of an inch.

Building Boldly for the Future

The plastic house of the future will probably bear little resemblance to today's home. Wood, brick and plaster may be replaced by polyester resins reinforced with fiber glass, combinations which can be molded to any shape and retain their structural strength without supporting framework. Plastic homes like the House of Tomorrow that was exhibited at Disneyland in California (left) can be built with synthetics, but mass production will have to wait until architects and builders grow accustomed to working with these materials.

Plastics already play an important role in conventional homes. Melamine resins, which can be used in molded or laminated form, have found wide acceptance because of their unusual strength and durability, in everything from dishes to kitchen counter tops. Vinyl and acrylic plastics are commonly used in floor coverings, pipes, paint and upholstery.

Public buildings often make spectacular use of plastics. The stained-glass windows of the Oklahoma church above are of the same hardy synthetic used to make canopies for fighter plane cockpits, brush handles, false teeth and translucent murals (page 27).

Taming the Rivers and Oceans

Plastics are supplying man with some unusual aids in his unrelenting struggle to control the waters of the world. When Los Angeles water officials decided they needed a dam across the Los Angeles River that could not only be adjusted for any depth up to six feet but could also be removed quickly for unhindered storm runoffs, plastics provided the means. A long tube of nylon-reinforced neoprene rubber was extended across the river *(left)*. It can be filled with air or water to any height, and is as sturdy as a fixed concrete dam. But in an emergency it can be emptied and flattened against the river bottom in 10 minutes.

Plastics may meet the even tougher challenge of preventing the erosion of sandy shorelines. Field tests with synthetic seaweed made of polypropylene—a plastic often used for luggage—have been conducted off the New Jersey and Virginia coasts and in the English Channel near Bournemouth. Fish took shelter in the seaweed as though it were real.

A COLLAPSIBLE DAM
A 150-foot plastic dam across the Los Angeles River, shown inflated to the five-foot level, can easily hold a man's weight although its neoprene skin is only an eighth of an inch thick. The dam, fastened to the channel's bottom and sides, holds 50,000 gallons of water when full.

SYNTHETICS UNDER THE SEA

Plastic seaweed, here being tested in a Philadelphia aquarium, is designed to protect seashores from tidal erosion and also provide sheltered feeding areas for fish. Experiments in the sea off the Atlantic coast of the United States and in the English Channel are being complemented by laboratory research to find the best arrangement in fighting the heavy surf.

A CAR'S BODY CAST OF PLASTIC
Plastic auto bodies are sturdy, rustproof and easily repaired. This sports car has a body cast from just two pieces of Cycolac (the same material used to make telephones). Other cars have been built successfully with bodies of various resins reinforced with fiber glass.

SYNTHETIC SKIS ON SYNTHETIC SNOW
Skiing down a plastic indoor practice slope, a model demonstrates that even man-made "snow" is possible. All of her clothing and equipment are plastic including the skis, which are epoxy fiber glass molded on a core of hickory and ash, and sealed with plastic wax.

The Changing World of Sports

EVERYTHING BUT THE GOLFER
Preparing to drive a synthetic rubber golf ball off a polyethylene green with a nylon-headed "wood," a Corfam-shoed golfer waits before teeing up. Plastics may not be able to improve his score, but at least his ball is cut-proof and his clubs are guaranteed to last for a lifetime.

A well-equipped fisherman, indulging in his favorite pastime on a Sunday morning, is a major beneficiary of the synthetics revolution. He is almost certain to have a molded fiber glass rod, nylon line and a tackle box of polystyrene. His plugs, floats, lures, net and boots are probably plastic; if he uses a boat, chances are that it will have a hull of plastic-impregnated fiber glass. The pleasure-boating industry used 220 million pounds of fiber glass in 1970.

Nowhere have plastics secured a greater hold than in sports and recreation. World records have been set by vaulters using fiber glass poles and by girl swimmers wearing four-ounce nylon suits (instead of a pound and a half of wet wool). Dacron sails, rugged and resilient, have not only propelled championship boats, but have also helped produce a remarkable upsurge in recreational sailing.

Plastic golf clubs, bowling balls and skis—both the snow and the water variety—are no longer novelties. But so far no plastic has been able to duplicate the wonderful crash and clatter of well-hit maple bowling pins.

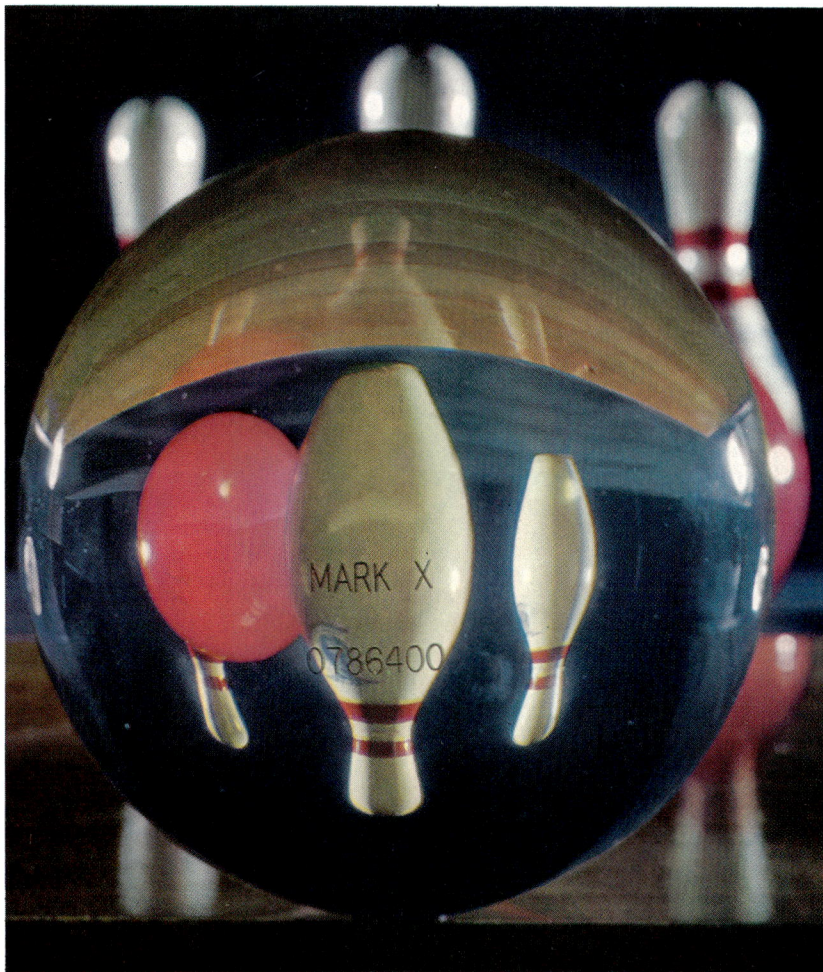

A NEW STYLE IN BOWLING BALLS
Colorful Lucite bowling balls, manufactured in seven translucent shades, are a far cry from the black vulcanized rubber ones which have been used in the sport since the late 1920s. Although manufacturers have thus far failed to produce a satisfactory all-plastic pin, they have been able to prolong the life of maple pins by sealing them inside shells of woven nylon.

Soaring in
a Synthetic Skin

The exceptional qualities of plastics have been utilized in a number of ingenious ways. The huge balloon on the right, for example, is a direct descendant of World War II's blimps. Like them, it is constructed of single-ply nylon fabric coated with neoprene. The combination weighs 4.04 ounces per square yard—about the same as fine batiste cloth. The figure of Bullwinkle Moose, one of around 100 made since 1924 for the annual Macy's Thanksgiving Day parade in New York, is 55 feet tall, contains 500 square yards of fabric and is floated by 7,500 cubic feet of helium.

While Bullwinkle entertains, other plastic balloons engage in vital work. Some carry scientific instruments aloft for weather and space research; another has been adapted by Oregon lumbermen to carry up to 10 tons of logs out of almost inaccessible timberland. One plan proposed for the raising of the sunken liner *Andrea Doria* involves another of plastic's qualities: buoyancy. Around 2,500 tons of liquid polyurethane foam would be pumped into the vessel. The plastic, it is hoped, would expand and harden quickly into a buoyant mass, displacing enough water to float the 29,082-ton ship to the surface.

A MOOSE ON THE LOOSE
Floating above Manhattan streets in the Macy's parade, Bullwinkle Moose's neoprene-coated nylon body is held together by 25 gallons of neoprene cement. The towering balloons used in this parade cost as much as $20,000 each.

A Striking New Medium in Art

Artists are constantly looking for new materials and techniques, and the plastics era has provided the combination of challenge and stimulation that great art thrives on. Sculptors have found more malleable materials, more workable "stone"; artists are making plastic mosaics and using paints that dry fast to a bright glow.

Sculptor Domenico Mortellito (at left), who has carved statues from rigid urethane foam since 1962, has found that it resembles Carrara marble but cuts like splinterless wood.

Artist Phyllis Stevens (bottom), working with Plexiglas, has developed a new art form with mosaic panels. Hung over or near windows, they assume different colors as the changing light diffuses through the plastic.

ARTISTRY IN ACRYLIC
The dazzling dragonfly above is the result of an all-plastic technique pioneered by artist Phyllis Stevens, shown at right in her studio. She glues bits of colored Plexiglas onto a sheet of clear Plexiglas with a special plastic adhesive. The finished panel is then sealed, baked and coated.

STATUES FROM FOAM
Sculpting in plastic, Domenico Mortellito uses urethane foam, a substance more familiar in sofa cushions or as spongy strips of insulation. The solid material, which will neither chip nor crack, can be delicately worked and then coated, painted or lacquered to a variety of finishes.

An Ever-blooming Garden

The art of making artificial flowers is not new; France's Madame de Pompadour is reputed to have had a garden of appropriately scented porcelain flowers two centuries ago. Glass, feathers and wax have also been used for floral reproductions. But not until polyethylene and polyvinyl chloride came into their own following World War II did it become possible to construct an entire garden of artificial blooms as perfect as the one at the right. This botanical surprise, which "grew" in the unlikely atmosphere of a Manhattan rooftop, contains some of the 500 varieties of plants reproduced by William Fuss, a German who helped introduce plastic flowers to the United States in 1947. Many people buy his expert copies of tropical or hard-to-grow plants to blend with real plants in their gardens.

The colorless, waxy polyethylene, widely used in manufacturing transparent plastic bags, is ideal for flowers: it holds its shape well, is flexible and can be delicately tinted. Fuss has made orange and lemon trees so realistic that birds build nests in them.

PARADISE IN POLYETHYLENE
This luxuriant garden will never need sun or water. Its birds of paradise, chrysanthemums, dahlias, anemones and even a wisteria tree are faithful copies in polyethylene. Owner William Fuss values the rooftop garden at $30,000. He "redecorates" it to match the changing season.

2

Discovering the Molecule

Hermes Trismegistus, god of magic, adorns a copy of a Ninth Century alchemy text.

TODAY'S HIGH SCHOOL STUDENT knows so much about the nature of matter that it is hard to realize what an impenetrable mystery it was as recently as a century ago. Man spent thousands of years blundering painfully toward an understanding of the materials which surrounded him and made his life possible. Until the 19th Century he understood almost nothing about the changes he himself effected—what really happened, for example, when he burned a log or etched a piece of metal with acid.

In ancient times men improved their lot by a trial-and-error chemistry. They learned empirically how to smelt ores into copper, iron and tin and to combine them into excellent alloys, how to make glass from minerals and wine from grapes, how to prepare various foods, medications and dyes. From these crude early processes to the controlled syntheses of today lay a long and tortuous labyrinth through which a strange assortment of men groped and argued, becoming lost in blind corridors, following baffling detours—and occasionally making brilliant, intuitive, correct turns that they often failed to recognize.

To start this journey the early chemists first had to determine which substances were the true elements of matter—those that could not be reduced to simpler substances—and to recognize which were compounds of elements bonded chemically, and which were merely unbonded mixtures. Then they had to create and gradually refine the concept of the atom as the smallest particle of an element, and later to visualize these atoms combining into structures called molecules. Finally they had to discover that, by rearranging many atoms of relatively few elements into different patterns, they themselves could create the giant, complex molecules that made up the substances of the natural, living world, as well as thousands of others that nature had never manufactured. The enormity of this task was staggering. For one thing, no one to this day has ever actually seen an atom, and only recently have the powerful electron microscopes pictured the actual structure of the largest giant molecules. The early, fundamental advances had to be made with crude instruments, by observation, deduction, guesswork and theories, some of them later proven uncannily accurate, but most of them found to be erroneous.

Indeed, chemistry might have saved itself centuries of frustration if the most basic of these inspired theories, advanced long before the birth of Christ, had not been lost in the babel of speculations of the ancient Greek philosophers. The word "atom" (from a Greek word meaning indivisible) was in fact coined by two of these philosophers, Leucippus and Democritus, to describe the tiny, unseen particles whose differing shapes, sizes and arrangements might give all matter its differing properties. But the mighty Aristotle favored a grander theory. The material world, he said, was composed of various combinations of the four indivisible "elements," water, air, fire and earth. It was this monumental error that was seized and elaborated upon for the next 2,000 years.

Even more mystical versions of elements and compounds dominated the dark alchemy of the Middle Ages, a long and futile search for the means to turn base metals into gold and to find an "elixir of life." Yet the alchemists at least left later chemists some useful tools: a supply of new acids, alcohols and alkalies, various chemical techniques such as distillation and crystallization, and a shelfful of flasks, retorts, crucibles and other instruments of the experimental laboratory. Toward the end of the 16th Century, mining and metallurgy made still other contributions: new metals and more accurate weights and balances. And the evolution of pharmacy added more knowledge of the behavior of chemicals.

But it was not until the latter half of the 18th Century—an age of enlightenment in all the sciences—that a series of true chemical experiments began slowly to unlock the basic secrets of the molecule. Some of these experiments were carried out by pharmacists and physicians seeking new remedies, but many of the most significant findings were made by talented amateurs—noblemen, clergymen, schoolteachers, an eccentric millionaire—to whom the new idea of chemistry proved both an intellectual stimulant and a fashionable hobby.

The gentlemen-scientists

Among the first of these gentlemen was a physician and teacher of chemistry in Edinburgh, Joseph Black, who in 1754 removed any lingering belief in the indivisibility of Aristotle's "elements" by showing that a "fixed air" (the gas carbon dioxide) could be released from a solid, earthy mineral (limestone) by heat or acid. Furthermore, he demonstrated that, after heating, the residue (lime), when left standing in air, could reabsorb the same gas and revert to limestone or chalk. Black had not only indicated that there was some carbon dioxide in the simple "element" air; he had also unwittingly split the molecule calcium carbonate ($CaCO_3$) into two smaller molecules (lime, CaO and carbon dioxide, CO_2) and then put them back together again.

About two decades later an English Unitarian minister, Joseph Priestley, broke apart another molecule by heating the red oxide of mercury, and obtained from it a most unusual "air" (oxygen) in which candles burned brightly and smoldering splints burst into flame. It was not long before air itself was shown to be a mixture of oxygen and other gases, notably nitrogen. Finally the wealthy eccentric Henry Cavendish, grandson of the second Duke of Devonshire, captured the still more spectacular "inflammable air" (hydrogen) released by metals treated with acid. When he combined it with Priestley's oxygen and introduced a spark, there followed a sudden, muffled explosion that produced several drops of pure water on the sides of the containing flask. This was the most startling development of all: not only was air composed of other elements, but now water was a compound of two gases! Cavendish had synthesized the molecule H_2O. Despite this progress, no one was yet able to visualize such elements as oxygen or hydrogen, or to deduce how they combined to form a compound like water.

But the day when these concepts could be grasped was close at hand. What remained to be done was the fitting together of two parts of an

A MEDIEVAL FURNACE was used for the distillation of common substances, a chemical technique that is part of a vast legacy left to modern science by the alchemists. While the apprentice in this 16th Century illustration might be collecting an essence of herbs, flies or oxblood to make medicine, a 20th Century chemist might use the process of distillation to purify materials for drugs and plastics.

intellectual jigsaw puzzle. In the 17th Century there had been renewed interest in Democritus' "atoms" when the English gentleman-chemist Robert Boyle had theorized that matter might be composed of many small, moving "corpuscles." In the same century, the physicist Sir Isaac Newton had speculated that matter was formed of "solid, massy, hard, impenetrable movable particles" that could not be broken into further pieces.

Assembling the atomic jigsaw

Now, at the dawn of the 19th Century, John Dalton, a quiet bachelor scientist from Manchester, England, put the puzzle together. He combined the ancient hypothesis of Democritus, the new evidence of Black, Priestley, Cavendish and others, plus his own observations to formulate the first practical theory of atoms and molecules, a cornerstone of modern chemistry. "The ultimate particles of all simple bodies are atoms incapable of further division," he wrote. "These atoms are all spheres, and are each of them possessed of particular weights, which may be denoted by numbers."

In teaching his atomic theory for 30 years, Dalton used little spherical symbols with different markings for the atoms of different elements, as well as diagrams that showed the spheres combining into various arrangements to indicate the structures of compounds. He also had a friend make a collection of little balls about an inch in diameter. They were painted a variety of colors to represent the different atoms, and as many as 12 holes were bored in each ball. In these holes he inserted wooden pins to join the balls together into crude models of molecules. Some chemists laughed at his "toys" and called his atomic theory a "tissue of absurdities." These cynics were unaware that Dalton had given chemistry its first accurate glimpse of molecular structure.

The day of the gentlemen amateurs was all but over. Dalton and his contemporaries were professional scientists who earned their livelihood as teachers of chemistry in universities. Their skill in research techniques was already impressive.

They were able to discover dozens of heretofore unknown elements, to manipulate hundreds of chemical reactions, and to set the stage for modern industry. Yet these men were working within a strictly limited range. They were dealing only with materials of the mineral, or nonliving, world: water, gases, metals, salts, certain acids, oxides. These inorganic substances were composed of small molecules, each containing a few atoms. Most of these small molecules were durable and stable. They could be broken into their elements, made into new compounds and returned to their original forms.

But the vital and useful products of the animal and vegetable world—wood, sugars, fats, oils—remained a baffling mystery. They burned easily and charred even when subjected to mild heat or certain acids, and what was left after a chemical treatment could never be put back together into the original. This gap in knowledge was about to close. The clearer understanding of mineral substances suggested ways of solving the complex mysteries of the animal and vegetable world. Already a few clues had been noted. By burning many products of the living world under a bell jar, the great French chemist Antoine Lavoisier discovered that they all yielded carbon dioxide and water, which indicated that they must all contain the elements carbon and hydrogen. Soon oxygen and nitrogen were also identified among the residues. As more and more compounds were tested, chemists came to a surprising but inescapable conclusion: despite their incredible diversity and complexity, all organic substances were composed principally of the same small group of elements: carbon, hydrogen, oxygen and nitrogen. Chemists had long been able to manipulate these elements in compounds of the nonliving world. Yet with all their knowledge and laboratory techniques they could not put the elements together into intricate products of the living world like wood or silk or sugar. Thus scientists were forced to believe that some mysterious "vital force" existed inside plants and animals to build up their complex and delicate chemical structures. Even the celebrated Swedish chemist Jöns Jakob Berzelius, who had divided chemistry into "organic" and "inorganic," spoke of the impassable gulf between the two.

Making synthetic urea

But in 1828 a young German chemist and former pupil of Berzelius, Friedrich Wöhler, bridged the gulf with a startling demonstration. As part of his studies of the cyanates, a toxic family of compounds, he heated a solution of silver cyanate and ammonium chloride, both members of the large family of "inorganic" substances. The ammonium cyanate formed by this combination was further heated and produced colorless, clear crystals which turned out to be identical in their chemical composition with the "organic" compound urea, a major waste product found in urine. "I must tell you," he excitedly wrote Berzelius, "that I can make urea without requiring a kidney or an animal, either man or dog."

Wöhler's methods were subjected to the inevitable questioning and skepticism that characterized chemistry. But he had indeed synthesized the first organic compound outside a living organism. As the notion of a vital force began its decline, other chemists were encouraged to attempt the construction of more organic compounds. Adolph Kolbe, a pupil of Wöhler's, synthesized acetic acid, previously found only in vinegar or the distillates of wood. Later, the French chemist Marcellin Berthelot took a still more dramatic step: he united acetic acid with several alcohols in the glycerol family to produce totally new fatlike substances. They were similar to natural fats but none exactly like them could be found in nature. Berthelot was among the first to show that chemistry could do more than imitate the products of living tissues: it could go on to create unique materials of its own.

But to understand the synthesis of organic compounds, chemists needed more refined theoretical knowledge. Probably the most important insight came from a strange coincidence. While Wöhler had been exploring the toxic cyanates, another young German chemist, Justus von Liebig, was studying the highly explosive fulminates, long used in the manufacture of firecrackers. Both men sent reports of their work to a chemical journal edited by the Frenchman Joseph Louis Gay-Lussac, who noted that the elemental formulas for silver cyanate and silver fulminate were identical (AgOCN) despite the fact that they acted very little alike. The phenomenon also came to the attention of Berzelius, who later discovered the same strange situation among several other organic compounds. Apparently an identical group of atoms of the same elements could form quite different chemical compounds.

To such compounds Berzelius gave the name "isomers," from the Greek for "equal parts." Wöhler's ammonium cyanate and urea, he explained, were isomers (with the common composition CH_4N_2O); heating had simply converted one to the other. As is now known, there are thousands of examples of isomers, including such unlikely pairs as vanilla flavoring and oil of wintergreen, and turpentine oil and limonene, the chief component of lemon peel. Clearly, the more atoms there are in a compound, the greater the number of possible combinations. For example, the compound $C_{10}H_{12}O_2$ can occur as oil of cloves, a mild antiseptic used by dentists, or as some 150 other substances.

Yet, though more and more examples of isomerism were discovered in the mid-19th Century, chemists had only a vague idea that they might have something to do with the arrangement of the atoms in the molecule. There was still no structural theory to explain the phenomenon.

One rational path through the maze was charted by Gay-Lussac who, while working with hydrogen cyanide (HCN), discovered that its unified carbon and nitrogen atoms could be shifted from compound to compound without breaking apart. This durable subsection of a molecule, in fact, acted very much like a single atom of a single element. Such groups of two or more atoms that remained faithfully together throughout chemical reactions were named "radicals," from the Latin word for "root," in the belief that they were the roots out of which more complex molecules could grow.

A new compound from smoky candles

But perhaps the most significant path was found by Jean Baptiste Dumas, a prominent professor of chemistry in Paris. His starting point was a seemingly trivial incident at a reception held in the 1830s by King Charles X in the elegant Palais des Tuileries. During the festivities, the tall candles around the room gave off a strange, heavy smoke and irritating fumes. When an official later asked Dumas what had caused this, he tested the candles in his laboratory and identified the fumes as hydrochloric-acid gas, which he traced to the chlorine used to bleach the candle wax. He found, moreover, that the chlorine was not merely present as an impurity but had unexpectedly joined with the wax, replacing some of its hydrogen atoms and forming a new compound with it. When the candles were lighted the combustion broke up the new compound,

A METICULOUS CHEMIST, Justus von Liebig won such 19th Century fame for his precise analyses of organic molecules that students from all over the world flocked to his laboratory in Germany. Liebig had intended to be a druggist, but switched to chemistry after he was fired for blowing out his employer's window with an experimental explosive.

uniting the chlorine atoms in it with the remaining hydrogen atoms to form the noxious gas.

Pursuing this lead, Dumas reacted chlorine with other organic substances, and discovered that the chlorine atom (as well as its chemical relatives, bromine and iodine) had the power to replace hydrogen in some compounds, atom for atom. He generalized this into the law of substitution vital to modern chemistry: in an organic compound the elements can be successively displaced and replaced by others. As later chemists found, substitution can produce prolific families of carbon compounds which, despite common parentage, are strikingly different from one another. One of the simplest hydrocarbons, for example, is methane, the principal compound found in the natural gas piped into thousands of modern homes for cooking and heating. Substituting an atom of chlorine for one of the four hydrogen atoms changes methane (CH_4) to chloromethane (CH_3Cl), sometimes used as a refrigerant, but almost useless for heating. Substituting three chlorine atoms produces trichloromethane—the anesthetic chloroform. When all four hydrogen atoms are replaced by chlorine, the highly inflammable methane finally becomes carbon tetrachloride (CCl_4), long used in extinguishers to put out fires, as well as to clean clothes.

Such phenomena were not yet clear to chemists of the mid-19th Century, but by now they did know enough to redefine organic chemistry as the chemistry of carbon compounds, and to realize that these compounds could be manipulated to produce others of fascinating complexity and untold uses. Because almost nothing was known about the internal structure of molecules, synthesis was based largely on crude methods of subtraction or addition of elements and radicals within known chemical formulas.

A chemist's brilliant failure

One of the most significant of the many attempted syntheses that followed resulted in a brilliant failure—so brilliant that it may be said to have started the modern synthetics industry by itself. In 1849, when malaria was killing thousands in the British colonies, it occurred to August von Hofmann, director of England's Royal College of Chemistry, that many lives could be saved by the artificial synthesis of the scarce antimalarial drug quinine, which came from the bark of the cinchona tree. Hofmann had been the first to perceive the chemical potentialities of coal tar, the black, sticky sludge that clogged the flues of gas works, and he set about separating it into its various constituents. One of these, a benzene derivative called toluidine, seemed to have a chemical constitution like that of quinine. An 18-year-old assistant of Hofmann's, William Henry Perkin, thought that by a series of substitutions he might transform the substance into quinine itself.

Perkin, the precocious son of a well-to-do builder, set about his ambitious task during his Easter vacation, in a rough laboratory he had fitted out in his father's house. But the series of reactions he caused produced

A LAMPOONED CHEMIST, Jean Baptiste Dumas, turns his back on the laboratory in this 1850 caricature. Dumas' radical—but correct—theories on the structure of organic compounds aroused such hostility from 19th Century scientists that in disgust he gave up a brilliant career as a theoretical chemist to become a French senator at the age of 52.

only a dirty, reddish-brown powder that settled to the bottom of the flask. More curious than dejected, he began again, starting with the simpler coal-tar derivative aniline (C_6H_7N). This time the final step, the addition of the oxidizing agent potassium dichromate, precipitated a mysterious black powder. Trying to find out what it was, Perkin dissolved the powder in alcohol. A stunning purple color appeared. Other chemists had thrown out such colored solutions as impurities. To Perkin, however, the brilliancy of his purple suggested a thoroughly regal dye. He tested it with some strips of silk. The color stuck fast; it would not wash out in soap and water as did the colors of most vegetable dyes, and strips he hung in the sun for a week did not fade.

The dye that produced the Mauve Decade

Perkin soon proved not only a persistent chemist but something of a businessman and engineer as well. At the suggestion of a friend, he sent off a bundle of his strips to a big dye house in Perth. Weeks passed. Finally came the reply from the canny Scots dyers: "If your discovery does not make the goods too expensive, it is decidedly one of the most valuable that has come out for a very long time." Determined to answer the "if," Perkin used his summer vacation and his older brother's help to set up a small-scale apparatus to test production problems and costs. Satisfied that his process was practical, Perkin took out British Patent No. 1984 and, much to the annoyance of Hofmann, quit the Royal College. Perkin then went to Perth to try out the dye in full-scale production. Unhappily, his aniline purple did not take evenly on yard goods of silk, and produced only weak lilac shades on cotton. But with the aid of the Scots he developed better fixing agents: a soap bath for silk, and tannin—the bark extract used in tanning leather—and a metallic oxide for cotton. Perkin then persuaded his brother and his contractor-father to help him build a chemical plant.

The little dye factory Perkin & Sons built during the summer of 1857 at Greenford Green, near Harrow, was to create the first industrial demand for the formerly wasted by-products of coal. But the crude benzene they bought was highly dangerous during the first manufacturing step, the addition of strong sulfuric acid and sodium nitrate. There were several explosions, fortunately without serious injury, until Perkin devised big, sealed cast-iron cylinders and learned to control his reactions properly. With these precautions he was at last able to transform commercial quantities of benzene into nitrobenzene into aniline into "mauve"—the color of the deep-lavender mallow blossom—a name given his dye by French dyers. Mauve, in turn, literally gave an era its name: the "Mauve Decade." The race was on. Other chemists began to unlock from coal a whole rainbow of synthetic, colorfast dyes, followed by a seemingly inexhaustible spectrum of drugs, flavors, detergents, photographic chemicals, plastics and explosives. But their fullest efforts awaited still more theoretical findings about the fundamentals of the giant molecule—most importantly, a clear visualization of its three-dimensional structure.

A FRENCH DYE FACTORY *(below)*, one of dozens that sprang up all over Europe after the synthesis of mauve in 1856, helped to meet a booming demand for a spectrum of new shades, most of them derived from coal tar. This sticky industrial waste—later the basic source of nylon and a host of modern synthetics—was distilled in huge flasks *(above)* to provide such popular 19th Century colors as magenta, Bismarck brown and imperial blue.

The Pioneers
of Chemistry

It has been only a little over two centuries since the infant science of chemistry took its first faltering steps toward the new era of man-made molecules. From the 17th Century on, unsung pioneers slowly began proving what had previously been only theory: that matter can contain elements or compounds; that compounds can be assembled to form new substances; that molecules can be juggled to reproduce exact combinations found in nature. The five men whose achievements are chronicled on these pages each contributed a base without which the molecule could never have been mastered —or its existence known. Their work was largely trial and error—four of the five stumbled on their discoveries almost by accident—and they all lacked the basic knowledge and apparatus available to any high school chemistry student today. But their findings in the years between 1754, when gas was proved obtainable from solids, and 1856, when the first chemical dye was synthesized *(right)*, were epochal. They not only helped make chemistry a full-fledged member of the family of sciences, but also laid the groundwork for one of the world's great industries—chemically produced synthetics.

MAUVE IN THE MAKING

A chemical solution changes from orange to black in one of a series of steps required to synthesize mauve, or aniline purple, from coal-tar derivatives. Discovered by William Perkin in 1856, mauve was the first commercially produced synthetic dye and marked the end of a century of experimentation with molecules, and the beginning of an age of dramatic achievement in synthetics.

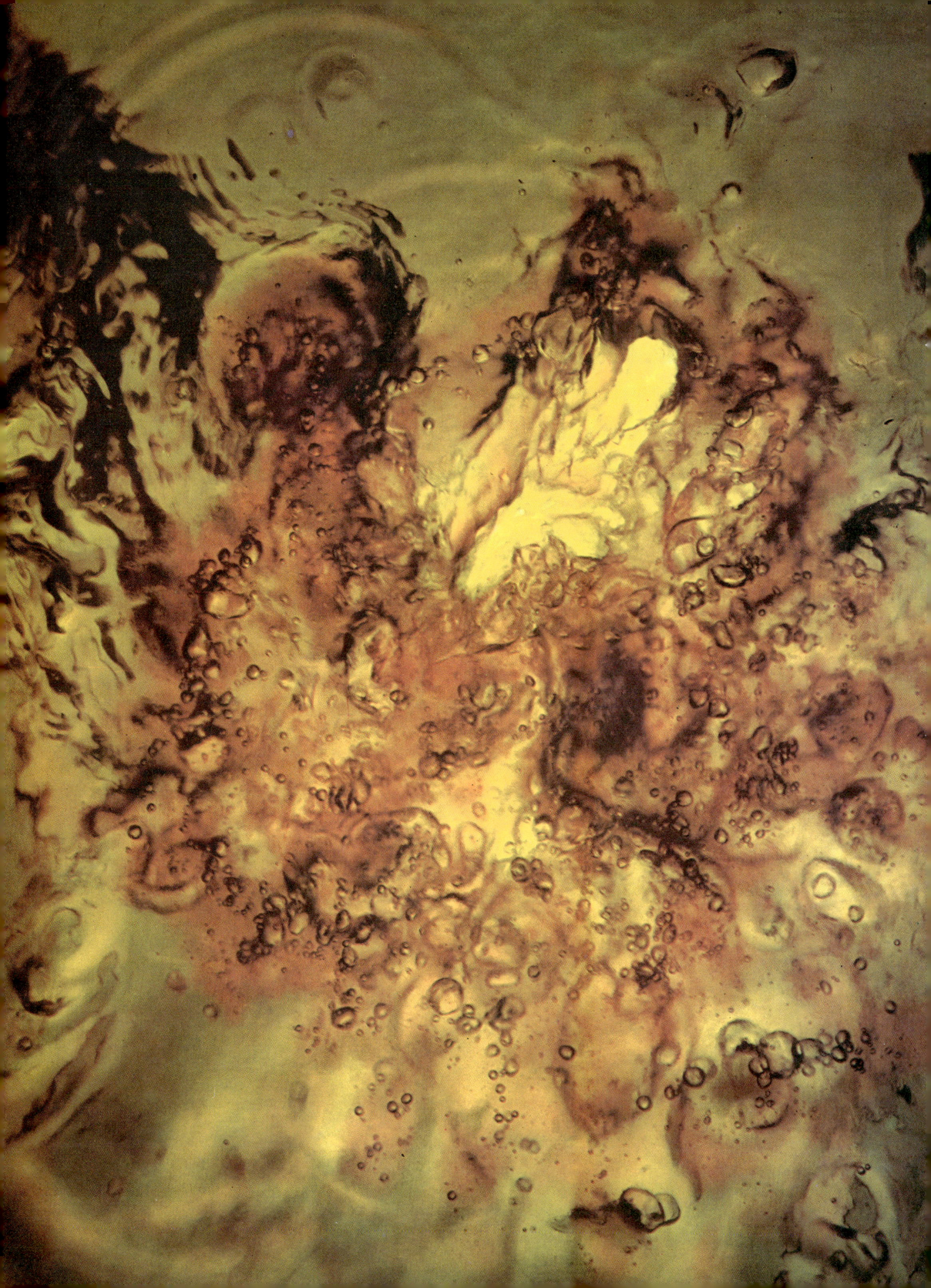

A DESTROYER OF MYTHS
Joseph Black, an 18th Century Scottish scholar, produced a gas—carbon dioxide—from a solid, and buried once and for all the ancient myth that all gases existed only in the atmosphere.

SIMPLE TOOLS OF DISCOVERY
Flasks, retorts and a balance, the apparatus of alchemy—a medieval science that sought to convert base metals into gold—were all that Black had to work with. He improvised ingen- iously to obtain accurate measurements, once noting that half the weight of a cubic inch of heated marble consisted of "as much air as would fill a vessel holding six wine gallons."

Getting Air from a Stone

The first man to break up a molecule and put it back together again was a 26-year-old Scottish doctor looking for a better way to dissolve kidney and gallstones. Joseph Black thought there had to be something less caustic than the lime-treated products being used in the middle of the 18th Century and more scientific than a popular remedy of the day that contained, among other things, powdered snails and eggshells, boiled herbs, soap and honey.

Black decided to see if he could produce a mild solvent by treating calcium carbonate substances—like limestone, chalk or marble—with acid. When he accordingly poured "spirit of salt"—hydrochloric acid—over marble *(left)*, he was intrigued by the fact that the dissolved lime that resulted was about three fourths the weight of the original marble. Weight reductions had previously been noted after burning and the old theory of alchemy linked them to the loss of a substance called phlogiston, the "essence of fire" that made burning possible. Now Black knew that acid, as well as fire, caused weight reduction, so the missing substance must be something else. He deduced that it was a gas, which he labeled "fixed air"—we now call it carbon dioxide. As proof, he mixed a mild alkali —a carbon dioxide source—with the residue of the marble. The original molecules were immediately restored —and so also was the missing weight.

A BREATH OF MARBLE
A cloud billows upward as hydrochloric acid is poured onto a marble slab, generating carbon dioxide that Black called "fixed air." By adding a mild alkali to the residue, Black was able to "refix the air," thus restoring both the marble's original weight and its chemical composition.

A CURIOSITY UNDER GLASS

As a concentrated beam of sunlight heats a basin of mercuric oxide, beads of mercury form and the candle in the bell jar burns with "a remarkably vigorous flame." Joseph Priestley, who observed this phenomenon in 1774, found that the heated oxide gave off four to five times its own volume in a strange kind of air which seemed to quicken fire. He called it "dephlogisted air"; we know it as oxygen. Priestley happened on his discovery by chance. Had he not placed a lighted candle in the jar during the experiment, he might never have noticed the gas.

The Minister's Superior Air

Joseph Black's discovery that a gas could be contained within a solid helped spur a wave of experimentation in the 18th Century, much of it by such talented amateurs as Joseph Priestley. An ordained Unitarian minister, he began dabbling in chemistry in 1767 when he was 34 years old. Five years later he perfected a method of charging water with carbon dioxide. The result was soda water, a beverage that became popular in England, both as a mixer for liquor and as a "preventive" for scurvy.

A favorite Priestley experimental technique was to heat various chemicals under glass by means of a lens (below) that concentrated the sun's rays, and then observe the different "airs," or gases, produced. It was by this means (left) that he trapped, in 1774, an air "five or six times better than common air for the purposes of respiration, inflammation and . . . every other use of common atmospherical air." Priestley had stumbled on one of the most important gases of all—oxygen. In similar fashion he investigated a number of other gases, including ammonia, sulfur dioxide and hydrogen chloride. Priestley left England in 1794 and spent the last 10 years of his life in the United States.

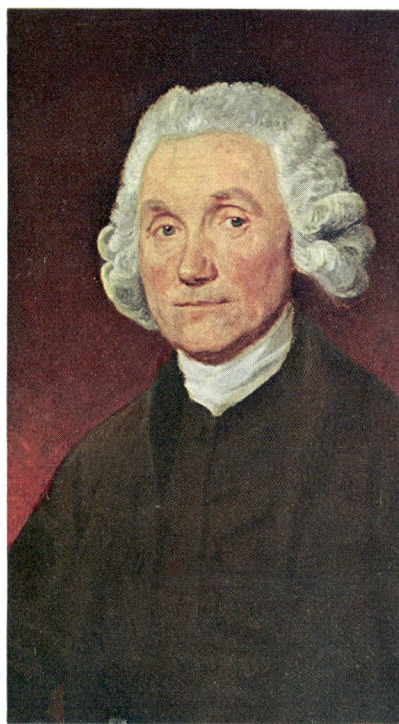

A DEDICATED AMATEUR
Joseph Priestley, like many 18th Century gentlemen, took up chemistry as a hobby. A chance meeting with Benjamin Franklin in London had sparked his interest in scientific experiments.

FOCUS FOR THE SUN
A burning lens—similar to this model, which Priestley later used in his Pennsylvania laboratory—concentrated beams of sunlight to heat the mercuric oxide in his famous experiment. Priestley lost much of his equipment when his English home was looted by a mob angered by his open sympathy for the French Revolution.

A Triumph and a Tragedy

The father of industrial chemistry was Nicholas Leblanc, a French physician who proved—perhaps a bit prematurely—the commercial value of chemistry. In 1775 the French Academy of Sciences offered a prize of 2,500 gold louis (about $13,000) to anyone who could find a cheap way to produce sodium carbonate—also known as washing soda or soda ash —used in making soap, glass and paper. Leblanc worked for 14 years to perfect a process requiring only salt, chalk, charcoal and sulfuric acid. He succeeded in producing soda but he failed to win the money. One version has it that the academy withheld the reward in the belief that Leblanc's process was not practical.

Leblanc borrowed 200,000 francs to open a plant in the Paris suburb of St. Denis in 1791. Within four years it was producing up to 700 pounds of soda daily. But Leblanc saw no profits after 1794, when the government confiscated the factory. It was returned—heavily damaged—by Napoleon in 1801, but Leblanc was unable to raise the money to rebuild it. Five years later, he committed suicide in the St. Denis poorhouse and was buried in an unmarked pauper's grave.

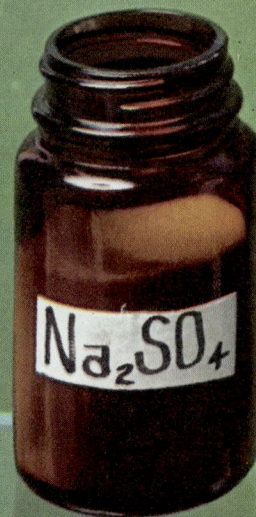

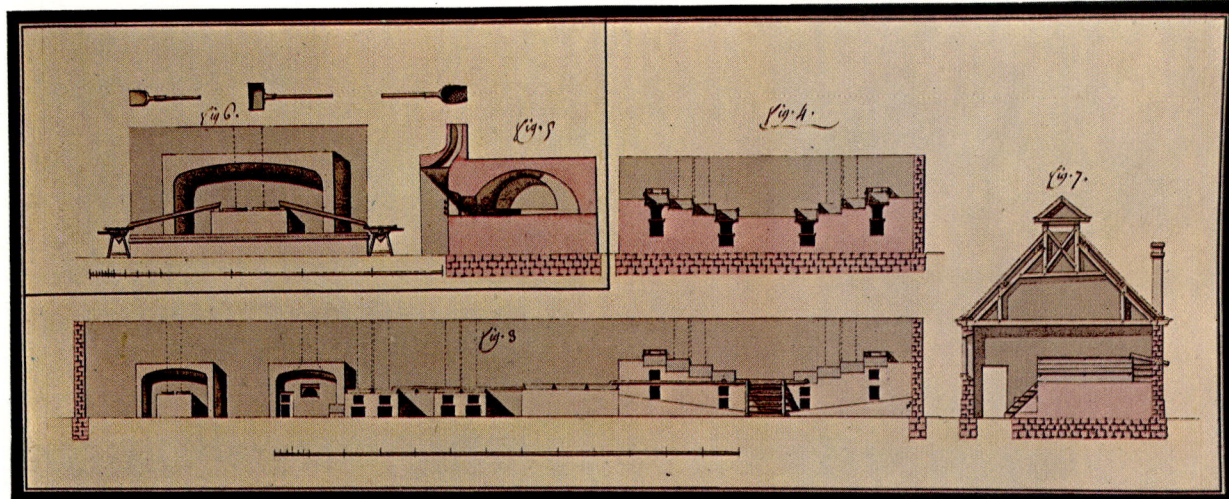

Nicholas Leblanc's statue in Paris.

Leblanc's pioneering soda plant, depicted in these schematic diagrams, covered about two and a half acres near the Seine River.

FROM SALT TO SODA

Leblanc's ingredients for producing soda are quite simple: sodium sulfate (Na_2SO_4), charcoal (carbon) and chalk ($CaCO_3$). Common salt is first heated with sulfuric acid to produce the sodium sulfate, being poured at right. This is heated with the carbon and the chalk (shown in beakers below their respective bottles) to form a mixture of sodium carbonate (soda) and calcium sulfide, to which water is added. Since only the soda dissolves, it can then be separated and evaporated to yield soda crystals.

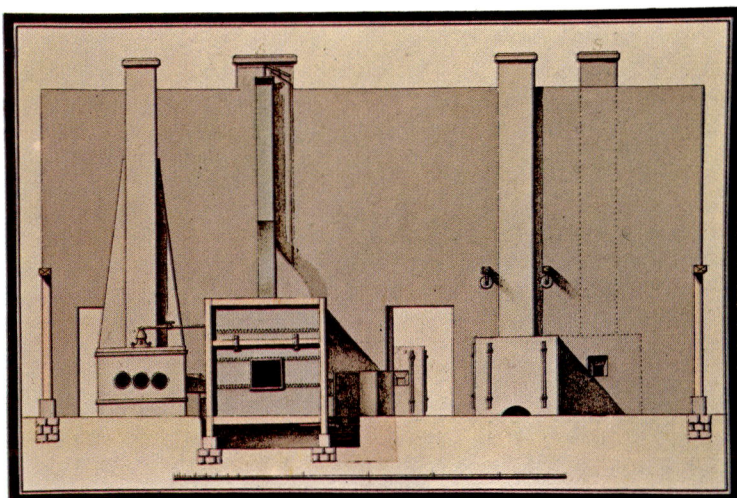

Heat-reflecting furnaces like these made the factory efficient and profitable.

A Synthetic Link between Worlds

When Friedrich Wöhler, a young German chemist, began a simple experiment in 1828, he never dreamed that he was about to synthesize a substance previously produced only by the body processes of a living creature. For all their hard-won scientific knowledge early in the 19th Century, the new breed of professional chemists clung to the belief that there was something holy about organic, or living, matter. It was supposed to contain a "vital force" beyond human knowledge. Inorganic matter might be freely investigated and even reproduced, but the mystery of organic matter defied challenge.

Wöhler had done considerable research on animal wastes. But in this particular experiment he intended merely to boil down a test tube of inorganic ammonium cyanate. When he examined the resulting crystals (opposite, lower right), he was amazed to find that they exactly duplicated the composition of urea, a substance in animal urine. Thus biology and chemistry were linked, breaking down a major obstacle to the synthetic reproduction of any molecule. Wöhler's discovery paved the way for chemistry's scientific coming-of-age.

A MODEL LABORATORY

Wöhler's laboratory, which can still be seen at the University of Göttingen, represented a great advance over the homely equipment of the 18th Century (page 40). Wöhler worked here from 1836 to 1882, constantly adding the finest apparatus available. The laboratory, equipped with a large reference library, was a model for all others in 19th Century universities.

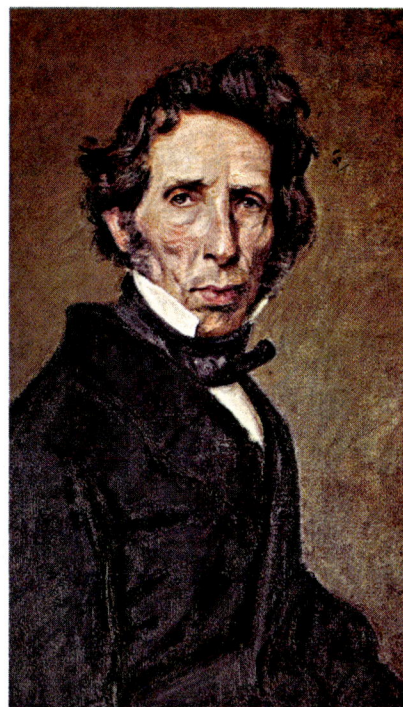

A DOUBTING DISCOVERER

Friedrich Wöhler, after synthesizing urea, repudiated the "monstrous and boundless thicket" of organic chemistry he helped create. He devoted the rest of his career to inorganic chemistry.

A CRUCIAL TRANSFORMATION

Silver cyanate spills into a beaker containing a solution of ammonium chloride in this restaging of Wöhler's classic experiment. The milky solution forming above the magnetic mixer contains silver chloride and ammonium cyanate—the simple inorganic compound Wöhler set out to study. Wöhler filtered out the silver chloride and evaporated the ammonium cyanate, expecting to find crystals of inorganic salt. However, when he examined the crystals (above) he realized they had the same composition as urea, an organic compound "impossible" to synthesize.

A Study in Purple

It was a failure by a persistent young chemistry student that led to the first commercially successful synthetic dye. William Perkin, only 18 years old, was having no luck trying to synthesize the drug quinine from coal-tar chemicals in 1856. In his little home laboratory, all he could get was a rust-colored solution that turned to black powder when he heated it, and then became a bright purple liquid when he dissolved it.

Perkin, charmed by the color, tested it as a dye and found that it resisted both sunshine and soap. He called it aniline purple, but it was as mauve that it began the color fad that led the last years of the 19th Century to be called the Mauve Decade—and that unleashed a torrent of synthetics over the next century.

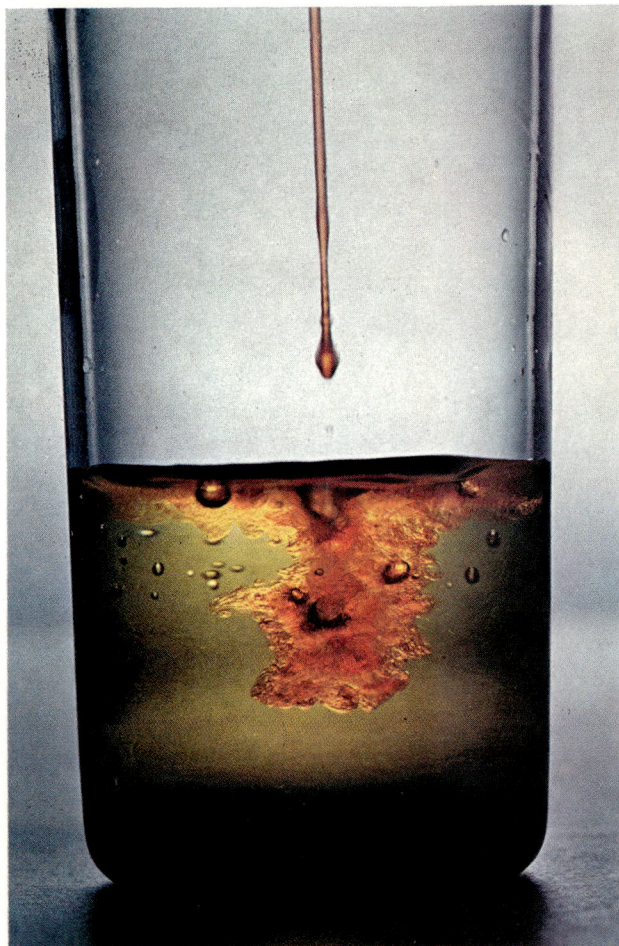

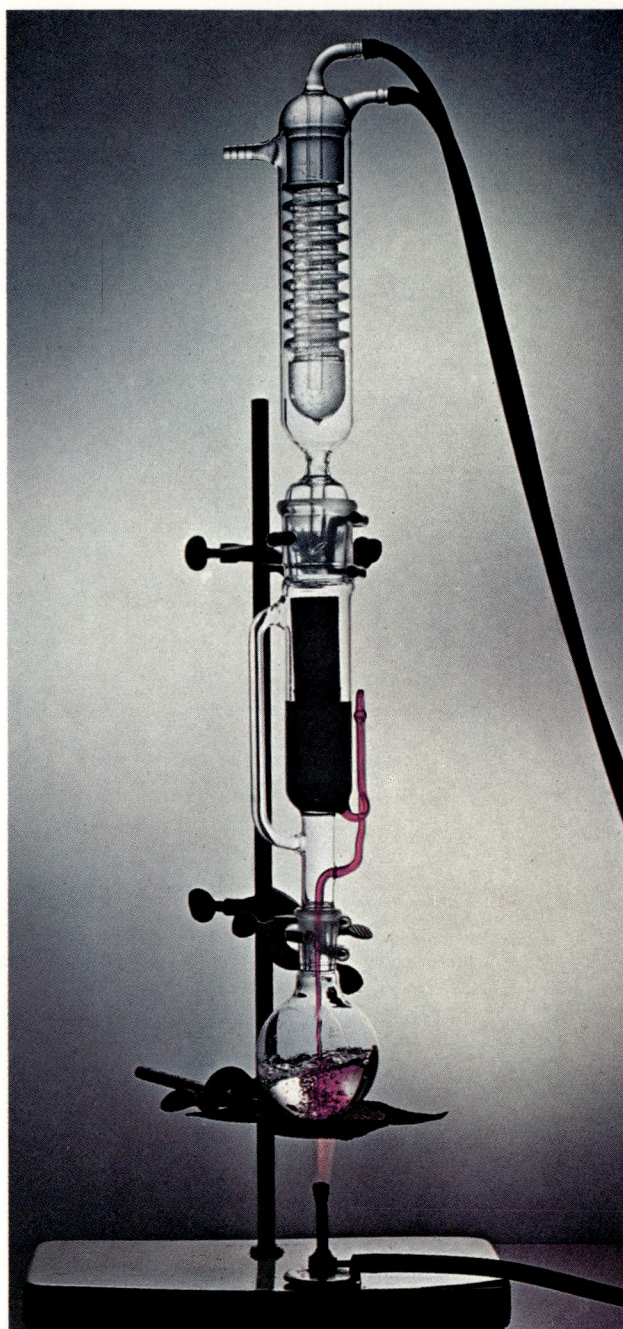

A DYE IS BORN
Perkin's mauve, here reproduced with modern equipment, is made by a process that begins with the mixing of orange potassium dichromate with aniline sulfate *(left)*. A black substance is formed, which is filtered, washed and dried. The powder is then placed in an extractor *(above)* and cleansed with naphtha to remove impurities, after which boiling wood alcohol is percolated through it to extract the dye, now a brilliant purple.

THE CREATOR OF A COLOR
Holding a skein of mauve-dyed silk, Sir William Perkin stands before his worktable. Perkin was knighted and this portrait painted in 1906, the 50th anniversary of his discovery of mauve.

3

A Blueprint for the Polymer

A microscopic view of groups of crystals is used by chemists to study polymer structure.

WHILE RIDING ACROSS LONDON on the top deck of a horse-drawn omnibus one balmy evening in the mid-1850s, Friedrich August Kekulé, a young chemist, fell asleep. "I began to dream," he reported later. "Atoms danced before my eyes. In the past, whenever these small bodies had appeared to me, they had always been in motion. Now, however, I saw how two smaller atoms often united to form a pair; how a larger one embraced two smaller ones; how still larger ones kept hold of three or even four of the smaller while the whole kept whirling in a giddy dance. I saw how the larger ones formed a chain, dragging the smaller ones after them. . . . The cry of the conductor, 'Clapham Road,' awakened me from my dream. . . ."

Thus, in a dream, was born one of the most fruitful concepts of modern science. Kekulé, a research assistant in his twenties, had visualized the basic structure of carbon compounds, the distinctive skeleton that gives these molecules their unique place in the world. He had perceived that they were chains, one carbon atom linked to another and that one to yet another, while atoms of other elements were hung onto the chain like a fringe. The picture he imagined was not completely accurate, but it was so close to reality that it has been used ever since as the working tool of organic chemistry. Besides being a brilliant speculation, it was amazingly useful: it showed how to match and even exceed nature with synthetics, and eventually it led to the far deeper understanding of giant molecules that 20th Century quantum mechanics theory was to bring.

For in the 1850s, despite some dazzling discoveries of the "new" organic chemistry—brilliant dyes, new medicines, perfumes, lubricants, solvents, and explosives unlocked from the blackness of coal tar and other carbon-compounded substances—there had been no inkling of how these wonders were put together. Chemists, for all their productiveness, knew little about the atomic architecture of the substances they worked with. The existence of atoms was accepted, and their combination into molecules was recognized. But beyond that lay mystery. How did the atoms of even the simplest organic molecule link up, and in what patterns? How did the patterns of atoms, and particularly of carbon atoms, relate to the properties of their compounds? Why, for example, did sodium join only one atom of chlorine to form table salt while carbon could take four atoms of chlorine to form the cleaning fluid now called carbon tetrachloride?

These were the key questions that Kekulé clarified, but the dream on the London bus was only the first of his several equally fruitful inspirations. It is possible that his brilliant success as a theorist of the structure of organic molecules may be explained by an educational detour he made in deference to his family's wishes. He was the son of a member of the Hessian defense ministry, and even in grade school had great interest in science. However, he also had considerable skill in drawing, and when he was ready for college, his parents insisted that he study architecture. But he soon switched to chemistry, despite the strong objections of his family. After receiving his degree, he worked in Switzerland for about a year and a half and then found a job as laboratory assistant in St. Bartholomew's Hospital in London. He was working there at the time he had his historic dream on the bus.

Late in 1855, he applied for the post of chemistry professor at the newly founded Zurich Polytechnic Institute but was turned down. The best job he could find was that of *Privatdozent* at the University of Heidelberg—an unsalaried lecturer who taught in his lodgings and swept out the parlor-laboratory after the students had left. His stepbrother, a merchant in London, supported Kekulé during this period.

But during his two years at Heidelberg, Kekulé worked out the practical details of his dream of the dance of the atoms. And in 1858, Germany's prestigious *Annals of Chemistry* published his classic article, "On the Constitution and Metamorphoses of Chemical Compounds and on the Chemical Nature of Carbon." Fame had arrived. He was sought after as a professor by a number of universities, won honor and respect as one of the leading philosophers of chemistry and, toward the end of his life, was even elevated to the Prussian nobility as Kekulé von Stradonitz; the knighthood was conferred on him by his former student, King Wilhelm II of Prussia.

Kekulé based his concept of a chain of carbon atoms hung with other atoms on the previously established facts of the combining power of atoms—or valence (from the Latin word for "power"). In 1852 Sir Edward Frankland, an English chemist, had pointed out that each kind of atom can combine with only so many other atoms. Hydrogen, for example, has a valence of one: one hydrogen atom can combine with no more than one other atom. Carbon has a valence of four, and each atom can connect to as many as four other atoms—including other carbon atoms. It is the way in which these connections are made that enables carbon to form very long chains of giant molecules.

Carbon's grasping arms

The carbon atom can be imagined as a small ball with four arms sticking out of it—north, south, east and west. Those arms represent carbon's valence bonds. Each one is a point of attachment; it can (and very readily does) grasp a valence arm of another atom. The simplest example is methane, sometimes called marsh gas because it bubbles up in swamps with other gases produced by decaying vegetation. Methane contains a single carbon atom surrounded by four hydrogen atoms. Each arm of the carbon atom grasps the arm of a hydrogen atom; since carbon has four arms and hydrogen one arm, the only combination possible is CH_4—one carbon atom attached to four hydrogen atoms.

The chain structure that distinguishes carbon compounds arises when

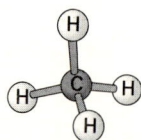

METHANE

THREE CARBON COMPOUNDS, methane, ethane and propane, are all components of natural gas. Carbon atoms build molecules by forming four bonds or links to other atoms. For example, methane, a simple carbon compound, has one carbon atom bonded to four of hydrogen. Carbon atoms can also bond to each other—ethane has two linked carbons with six hydrogens, and propane, burned in stoves, is a three-carbon-eight-hydrogen chain.

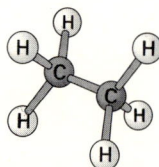

ETHANE

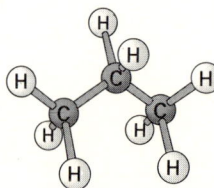

PROPANE

carbon atoms grasp arms with each other. One arm of a carbon atom can grasp the adjoining arm of its neighbor, and this process can continue indefinitely, producing molecular chains of two, three, four or many thousands of atoms. In each chain, the arms that are not linked to carbon neighbors are available to hold atoms of other elements. When just two carbon atoms grasp arms—eastward to westward so to speak—each atom has three other arms free. These may connect to hydrogen atoms, yielding the gas ethane—C_2H_6—the next larger relative of methane. When three carbon atoms link up, eight other arms remain free; if these are grasped by hydrogen atoms, the result is propane—C_3H_8—the cooking gas (shown below). And this process may proceed in this fashion to form innumerable carbon compounds.

Methane, ethane and propane are a family of fairly simple hydrocarbons—molecules made up only of carbon and hydrogen. But other organic families exist which include oxygen, nitrogen and many other elements as fringes on the fundamental carbon-hydrogen chain. Sometimes the chain itself includes other elements; sometimes the fringes include many other elements, not excepting carbon. But the structure is always basically the same: some arms of atoms link to form a chain while the remaining arms are free to grasp the fringe.

More bonds, less stability

In many compounds, the atom's attachment-arms double up. That is, two atoms may use two or even three arms to grasp each other. It would seem that these double and triple bonds ought to make the joint more stable. In reality the opposite is true. The most stable link between atoms is the single bond. Double and triple bonds are progressively less stable. The reason for this apparent contradiction lies in the nature of the carbon atom. It is most at ease when its four bonds project in four different directions. When two or three bonds are forced to line up between the same two carbon atoms, the additional bonds are strained and therefore more reactive. Other chemicals, heat or pressure cause these weaker links in the carbon chain to break readily.

Armed with the new and revealing theory of structure, 19th Century chemists renewed their attack on the molecular world; within a half century they added more than 150,000 new compounds to their shelves. As the work of analysis and synthesis proceeded, the torrent of new substances divided into two mainstreams.

One of these groups is called the fatty, or *aliphatic*, family (from the Greek word for "fat"); it includes such compounds as waxes, soaps, glycerin, lubricants, detergents and alcohols. Chemically, all these compounds are relatively straight chains of carbon atoms; they can be considered multiples and variations of one basic unit, methane CH_4, laid out in a series—carbon atom connecting to carbon atom.

The other mainstream of organic chemicals includes highly volatile and reactive substances which, because they are characterized by odors, are called the *aromatics*. Many of them were originally extracted from fragrant essences, spices and herbs. Oil of wintergreen, long used by pharmacists in ointments and flavorings, is a familiar example. The aromatics' readiness to combine with other chemicals makes them vastly more prolific than the aliphatics; Perkin's dye and those that followed—as well as most of the enormous range of modern medicinals, anesthetics and explosives—are partially aromatic.

A dream solution

To mid-19th Century chemists, the existence of these two types of compounds, one distinctively more lively and reactive than the other, was an intriguing puzzle. They suspected that the essential distinction lay in the structure of the molecules. But reliable clues were few. The aromatics, analyses showed, contained a greater proportion of carbon atoms; in fact, they always contained at least six carbon atoms—no compound with fewer than that number was aromatic. This confirmed the deduction that the basic unit making up aromatic molecules was benzene—C_6H_6. Why should a molecule composed of benzene be more reactive than one of methane? No special kind of bond connection could be invoked as an explanation. The distinction had to lie with the arrangement of atoms. The solution once more came from the clairvoyant Kekulé, and again in a dream. Here is the story as Kekulé told it in a speech in 1890: "There I sat, trying to work on my textbook but it did not go very well; my mind was elsewhere. I turned the chair toward the fireplace and dozed off. Again the atoms danced before my eyes. This time, the smaller groups remained modestly in the background. My inner eye, sharpened by similar visions I had had before, now distinguished bigger forms of manifold configurations. Long rows, more densely joined; everything in movement, contorting and turning like snakes. And behold, what was that? One of the snakes took hold of its own tail and whirled derisively before my eyes. I woke up as though I had been struck by lightning; again I spent the rest of the night working out the consequences of the hypothesis."

Kekulé's vision of a snake holding its tail led him to the concept of atoms linked in a ring-shaped benzene molecule. This idea enabled organic compounds to be grouped as open chain compounds (such as most aliphatics, made up of methane units) and ring, or cyclic compounds (such as the aromatics, made up of benzene units). Carbon's unique linking ability holds either type together; the almost infinite variety of natural and synthetic giant molecules is built up of carbon chains and carbon rings.

While the ring and chain concepts untangled most of the twisted threads of molecular structure, giving chemists surer control over their work, several balky problems remained. And they could not be solved

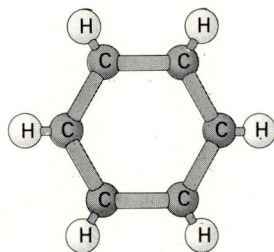

THE BENZENE MOLECULE above puzzled scientists for generations because they could not discern how it was bonded together. Friedrich August Kekulé supplied the decisive clue in 1864 when, in a dream, he saw a snake grabbing its own tail. Awakening, the chemist quickly translated the image of the snake into a six-sided benzene ring. This arrangement, it is now known, is established by the peculiar character of the bonds linking the carbon atoms. The bonds in benzene hold the carbon atoms closer together than do single bonds in compounds such as ethane, but keep them farther apart than do double bonds in compounds such as ethylene.

until a new view of the molecule—an image in perspective—revealed its shape in three dimensions.

One of the strengths of Kekulé's structure theories was their ability to explain the difference between isomers—compounds identical in atomic content but different in atomic layout, and therefore possessed of different properties. The existence of such isomers had been described by Louis Pasteur, the famed French chemist and microbiologist, who observed two different forms of tartaric acid in the laboratory and proved that their respective molecular structures were mirror images. But Kekulé's theories suddenly seemed to explain too much. They described isomers that did not exist. Once the layout of a compound had been deduced, it was simple enough to plot how sections could be switched around to create the various isomers. In many cases such pencil-and-paper conjectures were confirmed in the laboratory; the predicted isomers were found. But in other cases an isomer that seemed perfectly logical as a diagram could be neither discovered nor synthesized.

The classic instance was dichloromethane, a pungent-smelling liquid sometimes used in paint removers. It consists of a single carbon atom surrounded by two hydrogen and two chlorine—CCl_2H_2—atoms. Two possible arrangements of the atoms can be visualized: in one the chlorine atoms grasp the east and west arms of the carbon atom, while the hydrogen atoms hold the north and south arms. This symmetrical dichloromethane was thought to be the real liquid, quite easy to synthesize. But there should also be an asymmetrical isomer—with chlorine atoms grasping north and east arms while hydrogen atoms grasp south and west arms. It did not seem to exist nor could it be created.

The three-dimensional molecule

This unexplained gap in the rigorous structure theories was given an elegant solution about 1874 by Jacobus van't Hoff in Holland and Joseph Le Bel in France, who published their ideas only two months apart. Both van't Hoff and Le Bel saw that the problem of the nonexistent isomers disappeared if the molecule were viewed in perspective—as a three-dimensional object instead of a two-dimensional plan. Such a condition, postulated Le Bel, could be met if the carbon atom in "plane" form were lifted from the flat surface of the chemist's notepaper and "resolidified" into a pyramid with a triangular base. In the center of this pyramid hung the carbon atom, with its four bonding arms radiating out to the points of the pyramid.

Now it became clear why the isomer of dichloromethane was missing. It did not exist. Since the new three-dimensional carbon atom was symmetrical, it mattered not at all which of its arms were grasped by dichloromethane's two chlorine atoms—east-west or east-north—the isomers would be the same in the third dimension. Thus, when viewed in three dimensions, the two forms predicted by Kekulé's theories turned out to be the same form.

By extending this three-dimensional view, multiple bonding could also be explained. Single bonds between carbon atoms could be accounted for by placing the pyramids point-to-point; double bonds could be represented by placing them edge-to-edge, and triple bonds by placing them face-to-face.

These edifying and useful details of molecular structure had been painfully deduced from experiments with the simpler among the familiar organic substances—natural gas, alcohol, sugar, acetic acid (the main acid in vinegar), soap, oils. But most of the natural substances in everyday use—wood, paper, rubber, cotton, wool, leather, silk—are very much more complex in physical and chemical makeup. A chemist curious about their chemical composition could not learn much by applying the usual laboratory techniques—melting, volatilizing, or evaporating, distilling, dissolving—so revealing in the analysis of simpler substances. For example, leather will not melt, nor can it be easily dissolved; it reveals only a few of its chemical secrets by being volatilized, and only a few of those volatile components can be distilled.

Chemistry's nonidentical twins

As analytical techniques improved and research became more accurate in the latter part of the 19th Century, it became apparent that at least some of those complex substances were no different in chemical composition from some of the much more simple, more chemically vulnerable substances. The approximate formula for silk—$C_5H_8O_2N_2$—was found to be about the same as that of various compounds easily synthesized from organic acids and ammonia—but the synthetic equivalents had none of the valuable physical properties of natural silk. Atom by atom the two substances were chemical twins, but molecule by molecule they were obviously not the same.

Gradually it became apparent that there were a number of organic compounds like this—having the same general molecular formula but having entirely different characteristics. Cotton (cellulose) and glucosan (a simple, water-soluble sugar) both have the same formula: $C_6H_{10}O_5$. And yet they are entirely different. What caused that difference?

Much study and cogitation eventually revealed that there was only one possible explanation: the size and structure of the molecules themselves. Not only were these molecules much larger—giant molecules—but they were also built in a different way. They gained their size from a repetition of many simple units: one unit joined to another and that to still another in somewhat the way boards are nailed together to make a house.

All giant molecules, synthetic or natural, are formed in the same way, whether in the vat of a factory or in the body of a plant or animal: by the linking together of small molecules, or monomers (Greek for "one part"), into giant molecules, or polymers ("many parts"). Some polymers are

UNBONDED CARBON ATOM

A CARBON ATOM (above) can be visualized three-dimensionally as a tetrahedron with the nucleus (not shown) in the center of the pyramid. Each of the four apexes is the site of a bond (below). When the pyramids touch apex-to-apex, a single bond is created. A double bond forms when they meet edge-to-edge, a triple bond when they are face-to-face.

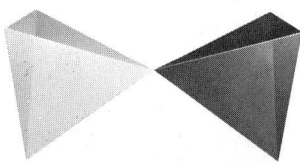

SINGLE BOND

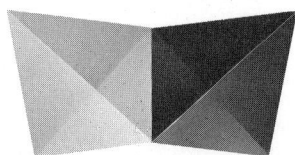

DOUBLE BOND

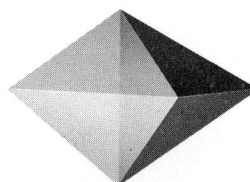

TRIPLE BOND

staccato repetitions of but a single kind of monomer; in other polymers, two or more monomers may alternate more or less regularly; the synthetic rubber used for tires is such a copolymer.

The structure theories of Kekulé explain how this linking together, or polymerization, of small molecules comes about. Since most polymers are strong materials, the bonds connecting the monomers must be stable —that is, single bonds. The single-bond connection can be made available for polymerization—an atomic arm set free to grasp another, so to speak— in either of two ways. In the simplest process, the free arm comes from the division of a double bond. Where two arms had linked atoms in the monomer molecule, one arm takes over that task while the other serves to link the monomer molecule into the polymer. In this addition polymerization, no atomic changes are made in the monomer molecules; each one keeps all the atoms it had and gains no new ones. It is simply connected by the arm of a single bond to its neighbor.

The other main process for linking monomers into a polymer provides the needed connecting bond by changing the monomer chemically. Some atoms are removed, setting free the bond that had formerly grasped them. The atoms usually involved in this condensation polymerization are hydrogen, chlorine, or small groups of hydrogen and oxygen. In many cases the monomer will lose two oxygen-hydrogen groups, retaining one of the oxygen atoms as a connecting bridge to the next monomer and turning loose the other atoms as a separate molecule of water (H_2O).

Small molecules sometimes polymerize into giant molecules spontaneously. That is what happens when perfume, left on the bathroom shelf too long, turns into a gooey mess—the terpenic monomers polymerize spontaneously into polyterpene, a resinlike substance. Even so, some outside urging is necessary to accelerate the bonding rearrangements—energy provided by light or heat, or oxygen in the air. In commercial syntheses, catalysts or initiators are added to start or speed the process, usually by forming chemical combinations with the ingredients.

The possibility of life in a laboratory

Today these steps in the creation of a new material can be planned in advance, and controlled as they progress, with the artistry that a composer exercises in scoring a symphony. In the 18th Century chemists worked by feel, knowing little of what happened in their retorts and understanding less. But with the guidelines provided by Kekulé, Pasteur, van't Hoff and Le Bel they quickly moved beyond Perkin's accidental synthesis of a dazzling dye. First came the alteration of the giant molecules of nature, then the creation of polymers that were entirely man-made, next the production of molecules to order—and eventually, perhaps, will come the synthesis of substances that are alive.

Forging the
Giant Molecule

The twisted, bumpy chain snaking across the opposite page is a tiny bit of clothing. If a woman's stocking could be magnified a few billion times, an endless array of such structures would emerge, for what the picture shows is a model of a small portion of a single nylon molecule. Devised and constructed by man to fill a specific need, it has properties unmatched by any molecule found in nature. Yet it began, as does every synthetic, with nature's own building blocks, the small, simple molecules called monomers that are derived from such sources as petroleum and coal.

To build a giant molecule, or polymer, like the one shown here, the chemist must begin with one or more kinds of these building blocks, altering their structures if necessary, and then stringing together as many as several thousand of them into long chains. By choosing appropriate components and controlling their arrangement, he can get such desired qualities as strength, durability, elasticity and insolubility. In effect, the chemist's ever-increasing skill enables him to stack tiny building blocks in a multitude of patterns to produce virtually any substance his fertile imagination can create.

A LONG LOOK AT NYLON
A model of a nylon polymer, its approximately 45 linked units containing about 1,700 atoms, stretches into the distance. The backbone is carbon (black), with every seventh atom nitrogen (blue). Along the side are hydrogen (white) and oxygen (red). A giant among molecules, the polymer is 84 billionths of an inch long—invisible even through the most powerful microscope.

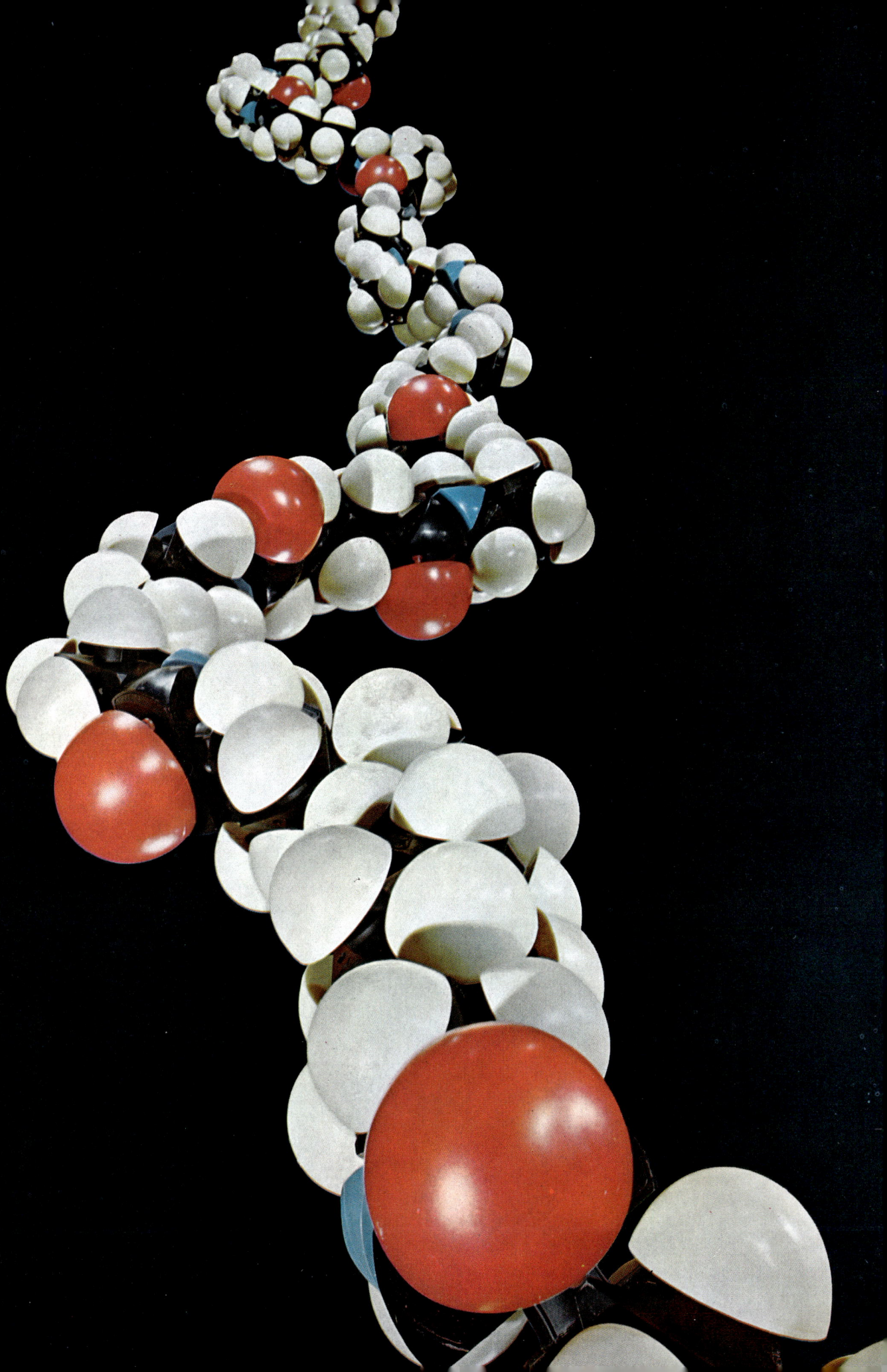

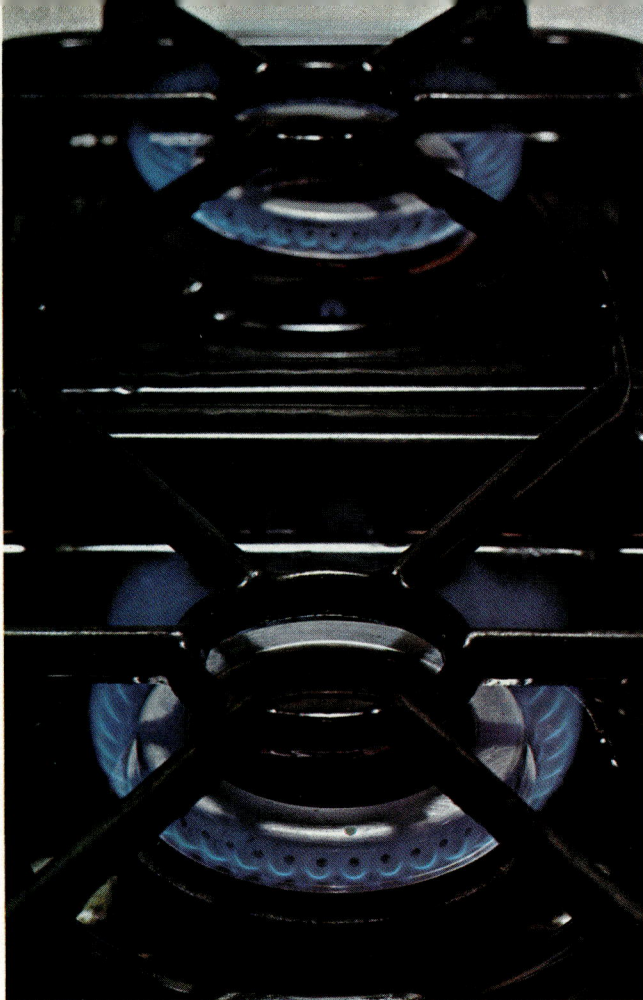

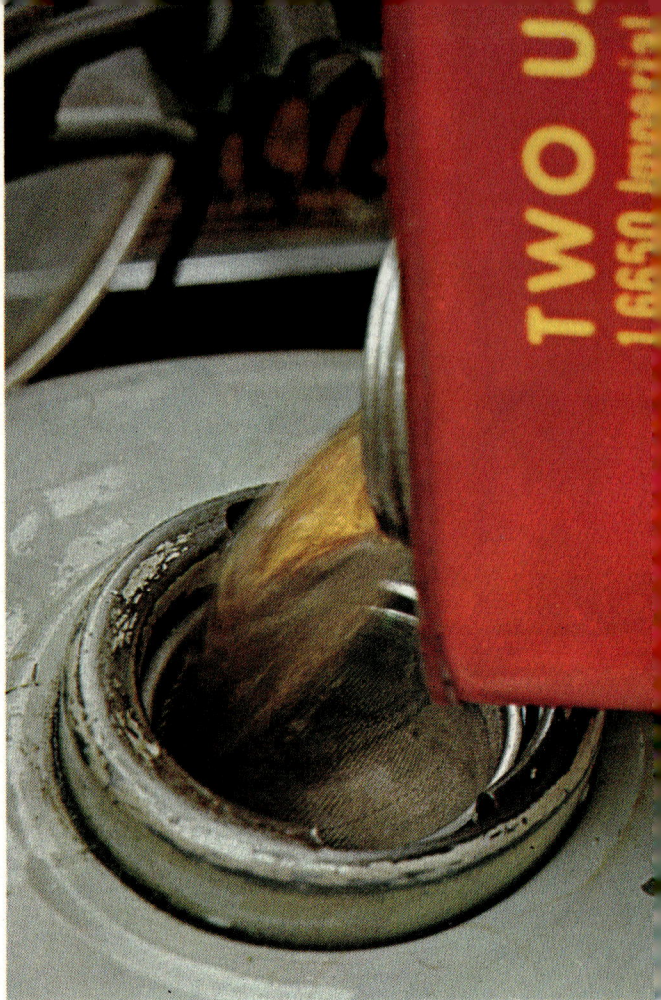

GASES: METHANE TO BUTANE

Nature's methane series, as grouped by modern chemistry, begins with the smallest and lightest hydrocarbons. Methane and its next-larger relative, ethane, make up as much as 90 per cent of natural gas *(above)*. Three-carbon propane is bottled gas, often used as a refrigerant and as raw material to make other chemicals. The largest and heaviest of the methane gases are the four-carbon butanes, used in cigarette lighters. Propane and butane can be turned into liquids under pressure, and together make up the widely used liquefied petroleum fuel, LP-gas.

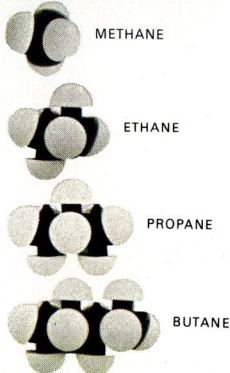

METHANE

ETHANE

PROPANE

BUTANE

LIQUIDS: PENTANE TO HEPTADECANE

When the growing methane chain takes in five carbon atoms, liquid pentanes—used in anesthetics and to make artificial ice—are formed. As the molecules get larger, the liquids they form become heavier. In the process they create a number of different oily liquids including heptanes (seven carbon atoms) and octanes (eight carbon atoms), some of which are used to rate the performance of gasolines. The larger hydrocarbon molecules form liquids as heavy as the heptadecanes—semisolids 17 carbon atoms long that are sometimes found in crude oil.

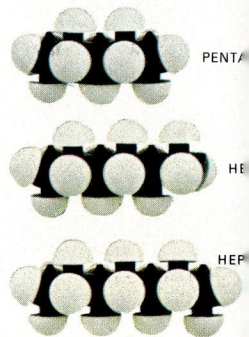

PENTA

HE

HEP

The paraffin molecule shown in this model is basically a stretched-out methane molecule. Its 30 carbon atoms and 62 hydrogen atoms give

Nature Builds
a Molecule Chain

In structuring small molecules into synthetic polymers, the chemist often is merely copying a process older than life itself. For the molecular chain exists all around us in nature. In the examples shown here, a single natural molecule—methane—provides the basic unit for all three types of matter: gas, liquid and solid.

The methane molecule, a hydrocarbon originally produced by the decomposition of organic matter, consists of a single carbon atom and four hydrogen atoms. In this pattern, it makes up the highly inflammable firedamp, a gas dreaded by miners. The next-larger hydrocarbon, ethane, has two carbon and six hydrogen atoms; both compounds are found in pockets of natural gas. The addition of one or two more carbon atoms, with their attendant hydrogen atoms, creates heavier gases, but when still more carbon atoms are joined, liquids are formed. As soon as the carbon count reaches 18—i.e., 18 methane units connected together—the result is a solid paraffin.

The 30-carbon-atom paraffin shown below is one of the larger of the methane series ordinarily found in nature. Today chemists carry this repetition of the methane unit much further, stretching the series to include polyethylene, a polymer containing up to 200,000 carbons within its long chain.

SOLIDS: THE WAXES

When 18 or more carbon atoms and their satellite hydrogen atoms are linked in the methane series, they produce a group of solids known as petroleum waxes. The shorter chains, which contain up to 30 carbon atoms, form the common paraffin waxes used to seal jars of jelly *(above)*, make candles, coat milk cartons and weatherproof tarpaulins. From the longer chains, which may be stretched to include 70 carbon atoms, are derived the higher-melting microcrystalline waxes that serve as a base for such products as petroleum jelly and shoe polish.

it a weight and stability lacking in the lighter gases and liquids, and impart the waxy characteristics of all solids in the methane series.

ETHANE—THE SINGLE BOND

Model molecules of ethane, above and in a diagrammatic view at left, show how each carbon atom uses three of its four bonds to hold hydrogen atoms, leaving only one to form their carbon-to-carbon bond. Although this bond can be broken by burning—as a component of natural gas, ethane makes a good fuel—the molecule will then completely disintegrate, and cannot be reassembled to make a polymer chain.

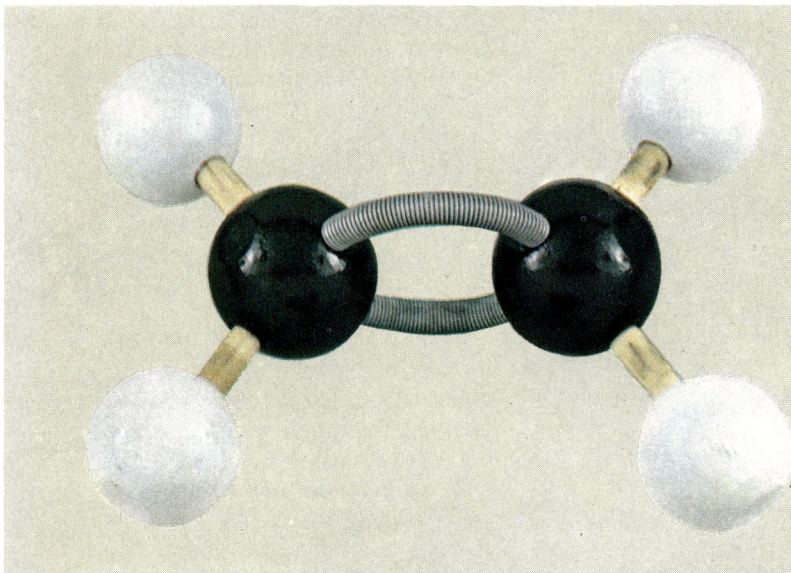

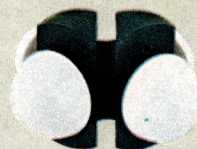

ETHYLENE—THE DOUBLE BOND

Four atoms of hydrogen and two of carbon team up to form ethylene, whose carbon atoms are linked by a double bond. If the weaker of the two bonds is broken while the stronger one is retained, the bond thus set free can link ethylene intact to another molecule. Chemists have found ethylene—produced from natural gas or petroleum—handy for a host of synthetics, including such familiar ones as saran and Dacron.

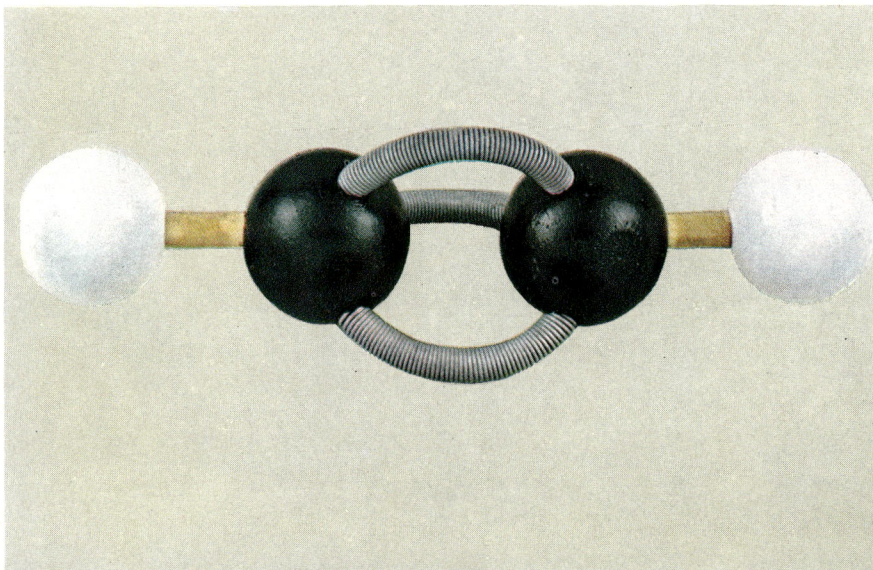

ACETYLENE—THE TRIPLE BOND

The most explosive hydrocarbon, acetylene, is formed by a triple bonding of two carbon atoms. Although acetylene reacts to other compounds unpredictably, and often violently, it is nonetheless valuable as a building block for acrylic fibers and synthetic rubber. Acetylene compounds are also commonly used as fuel in metal-cutting torches (right), because of the tremendous heat that is generated when they burn.

Three Types of Atomic "Glue"

The force that holds most giant molecules together is the bonding of carbon atoms—i.e., their ability to form bonds with each other and with other atoms. The number of carbon-to-carbon bonds may determine if a molecule can build a polymer chain.

In hydrocarbons, the atoms of carbon build more bonds between themselves as the number of hydrogen atoms decreases (*opposite*). In ethane (*top*), each carbon atom holds three hydrogen atoms as well as its partner carbon. Unless an atom is removed, there are no bonds free to join in a polymer chain. The ethylene molecule (*center*), with two fewer hydrogens, has two bonds between carbon atoms. One of these can be readily "untied" to link the molecule to a chain. As for acetylene, its triple bond is so easily broken (*below*) that it flies apart violently, which severely limits its use as a polymer builder.

WHITE HEAT FROM HYDROCARBONS
A triple-bonded gas—methylacetylene—mixed with pure oxygen and ignited, turns steel from solid to liquid in the white glow of its 5,500° F. heat. Pure acetylene burns even hotter, but its explosive nature makes it difficult to transport as well as extremely dangerous to handle.

Two Ways of Adding Molecules

The two plastic bottles at left were manufactured of the same substance: polyethylene. Yet, when both were subjected to the same amount of heat for the same length of time, one bottle melted, the other did not.

This seeming paradox can be explained by the manner in which their molecular chains—each made by linking thousands of ethylene monomers—were assembled. The unmelted bottle has its molecules in unbranched chains. Such chains, crowded close together, give this type of polyethylene a rigid and tough quality, making it an ideal container for liquids, among many other uses. The melted bottle contains chemically identical ethylene monomers, but arranged in branched chains resembling a set of dominoes. The result is a much looser and less sturdy polyethylene, but one that is softer and more flexible—well suited for the wide variety of squeeze bottles that can dispense everything from shampoo to catsup.

Both polyethylenes are examples of addition polymers—giant molecules requiring only the simple and repeated connecting of monomers.

THE OVEN TEST
Two polyethylene bottles, heated in a 240° F. oven, illustrate the difference in strength that can be built into one basic plastic. The melted bottle is composed of branched polyethylene molecules, while the intact container is made from more densely packed, unbranched chains.

AN UNBRANCHED CHAIN

The ethylene molecule at right is the basic building block of all polyethylene plastics. The sequence below illustrates how the individual molecules react when a catalyst—chromium oxide—is added and the mixture heated. The molecules quickly become attracted to one another and join to form one continuous chain. When all have been incorporated, the result is a single giant polyethylene molecule *(bottom)*.

A BRANCHED CHAIN

Ethylene molecules sometimes form a chain with irregular branches. This reaction takes place when the ethylene is heated to 480° F. while subjected to a pressure of 30,000 pounds per square inch. Since the branches prevent the giant molecules from lining up in close-packed rows, the resulting plastic is less dense and therefore more flexible—but also more susceptible to heat—than the unbranched chain type.

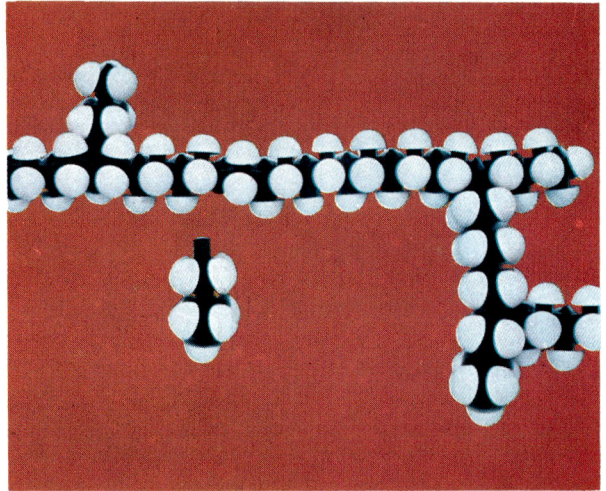

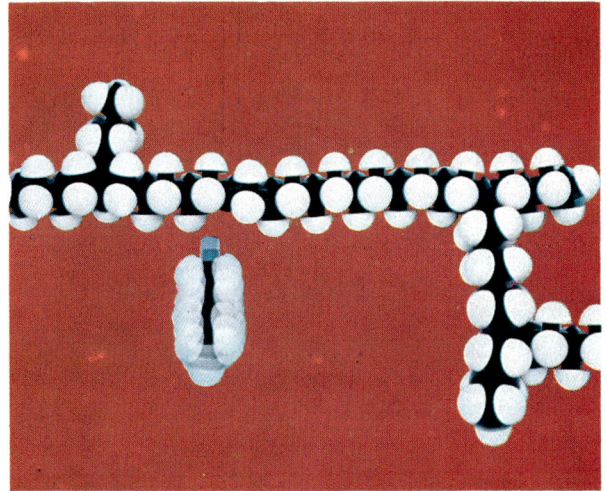

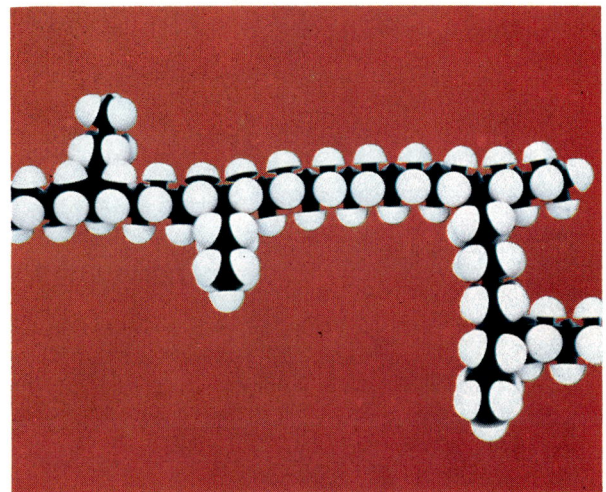

MOLECULES

MIXTURE

REACTION

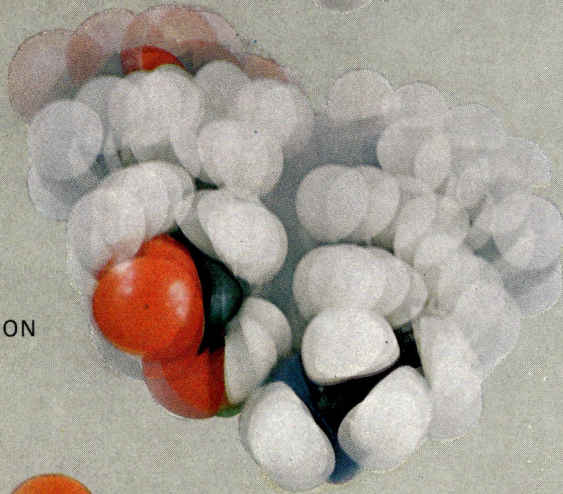

CONDENSATION

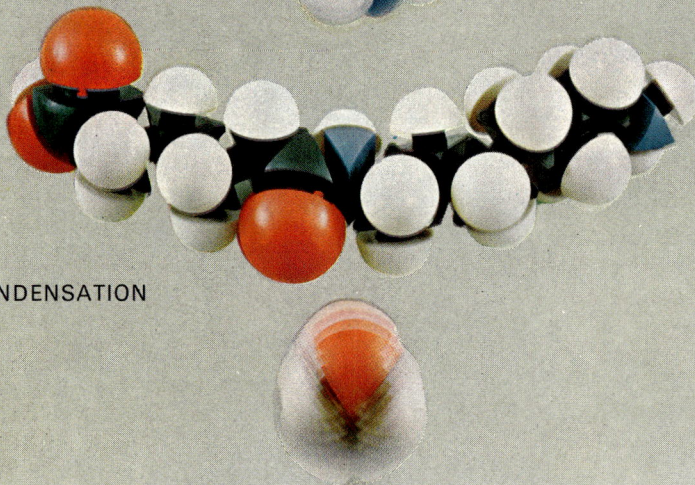

THE RECIPE FOR NYLON

The condensation process by which nylon is made begins with two different compounds: one, a molecule of adipic acid *(far left),* contains carbon, hydrogen and oxygen; the other, a molecule of hexamethylenediamine, is a nitrogen-containing compound. These compounds adhere to each other and become a larger molecule called a nylon salt. When the salt is melted in a vacuum, a more pronounced chemical reaction takes place. The compounds link chemically, in the process releasing hydrogen and oxygen atoms as water (H_2O), and the reassembled molecule *(bottom)* is repeated some 45 times to form the elongated nylon polymer.

Nylon's orderly structure, as seen in this molecula

Assembling the Nylon Polymer

While the relatively uncomplicated process of addition can be used to build a giant molecule from the simple double-bonded monomers, a more complex chemical reaction, known as condensation, takes place when other types of monomers are involved.

In the manufacture of nylon, the elements of an acid and an alkaline molecule must be combined. The polymer chemist employs a two-stage process to achieve this union. He first blends the two molecules into a single molecule of organic salt, then melts the salt by heating it in a vacuum at over 500° F. This produces a new condensed molecule containing all of the original atoms, less three—the two hydrogen atoms and one oxygen atom that constitute water. What remains is a unit called an amide (lower left), which then links with about 45 identical units, completing the long nylon polymer, or polyamide.

model, is strengthened by invisible links between oxygen atoms (red) and hydrogen-nitrogen groups (white and blue) on adjacent chains.

Stringing Molecules Together

Today's polymer chemist is, in effect, a molecular engineer who uses his knowledge of chemistry to create made-to-order synthetics. With a batch of basic molecules as his raw material, he can produce substances that are supple or stiff, fireproof or meltable, bouncy or sticky.

To obtain the properties he desires, the chemist takes into account three key factors: the chemical composition of his building blocks; the possible shapes of the polymer chains they can produce; and the alignment of these chains within the final product. Some of the possible variations are shown at right. Different combinations of these three factors can produce synthetic products as unalike as nylon stockings and vinyl flooring. Ordinarily the polymer chemist seeks out the best synthetic for a specific product. For example, a substance for a child's toy airplane should be strong enough to hold its shape, yet flexible enough to withstand the inevitable crashes. The structuring of benzene and ethylene molecules at lower right provides an ideal substance: a polystyrene-polyethylene mixture.

How the molecular arrangements shown here have been translated into a useful variety of synthetic products is illustrated on the following pages.

FLEXIBLE

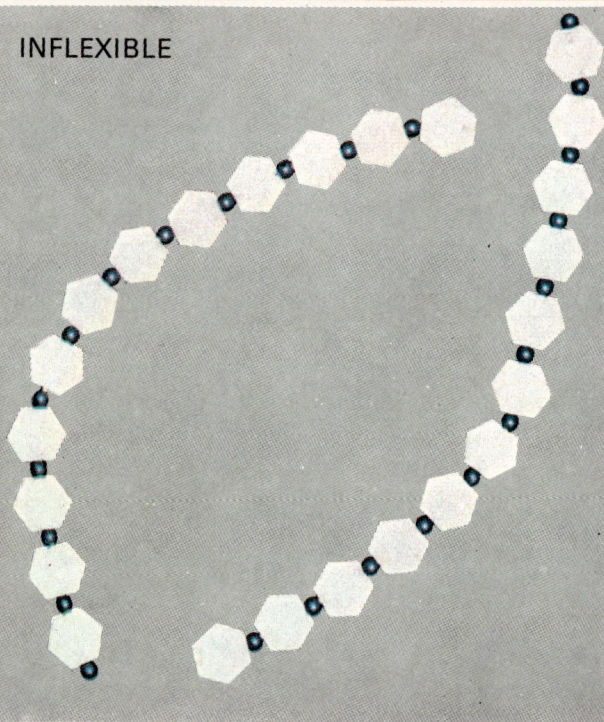

INFLEXIBLE

BENDING THE POLYMER
The shape of the polymer chain varies according to the size of the molecules used. The free-flowing serpentine *(top)*, made up of small molecules, lends flexibility to both the polymer and the resulting products—like polyethylene bags. There is more inherent rigidity in the arrangement below, a grouping of large benzene molecules. This is typical of such plastics as Merlon and Lexan, used for space-helmet visors and aircraft gauges.

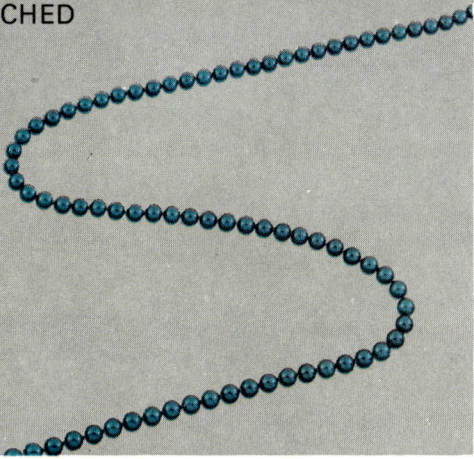

UNBRANCHED

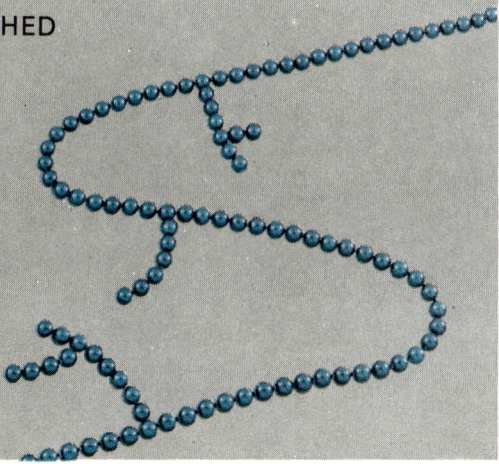

BRANCHED

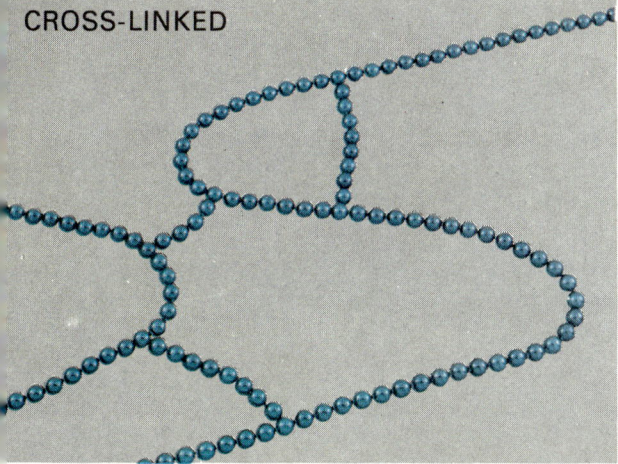

CROSS-LINKED

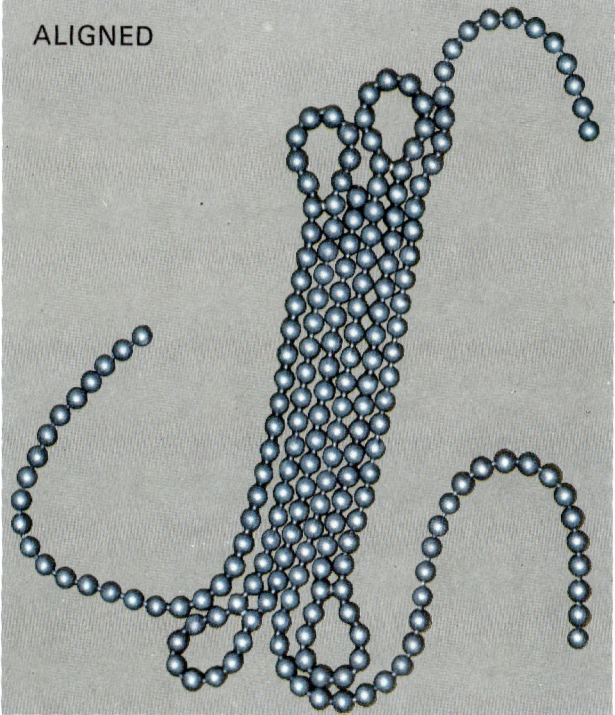

ALIGNED

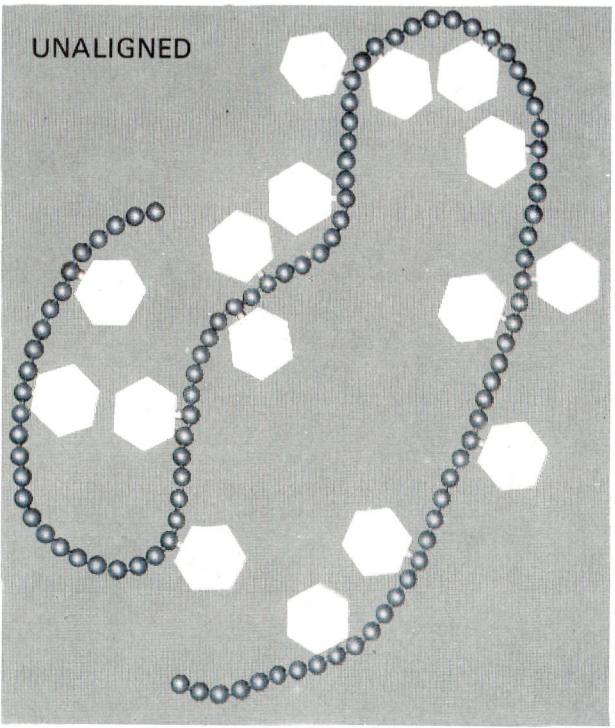

UNALIGNED

VARIATIONS ON A CHAIN

A simple polymer chain takes on different characteristics when branches or cross-links are added to its structure. As a result, the unbranched chain of high-density polyethylene *(top)*, which has a melting point of 275° F., becomes low-density polyethylene, with a slightly lower melting point, when branches are added *(center)*. But when cross-links are introduced *(below)*, the polymer is at its strongest and resists melting up to 400° F.

A DIFFERENCE IN DENSITY

The alignment of polymers within a substance has a direct bearing on the density of the final product. Chains which can align themselves closely, like those of high-density polyethylene *(top)*, tend to produce stiff and heat-resistant synthetics. When large benzene molecules are added to make the polystyrene-polyethylene mixture *(below)*, the chains can no longer align tightly. The result is a less dense and more flexible plastic.

Polymer Possibilities

Anyone who repairs things with rubber cement, rides on synthetic tires, combs his hair with a plastic comb, or cooks frozen food without taking it out of the bag is sampling the art of the polymer chemist. The essential qualities of each of these products—stickiness, resiliency, glassiness and durability—were built in by the chemist, using the techniques of polymer-making shown on the previous pages.

It is the particular combination of factors—as much as the building blocks used—that makes these products so different. As indicated in the color-coded blocks, both rubber cement and synthetic rubber have flexible, unaligned polymers, giving both of them the desired qualities of elasticity. But the cement's shorter, unbranched molecules are far less rigid than the tire's longer, cross-linked molecules, which were designed to resist curb-pounding and road friction.

FLEXIBLE	UNBRANCHED	UNALIGNED

FLEXIBLE	CROSS-LINKED	UNALIGNED

The glasslike acrylic comb and the sturdy polyester bag also share two properties—each has inflexible, unbranched chain molecules. But the molecules of the comb are unaligned, while those of the bag are aligned. It might seem, therefore, that the bag should be less flexible than the comb, but there is no paradox here—it is the thickness of the two products, not their molecular makeup, that makes the difference. For if the comb's acrylic plastic were rolled into a film, it would be more flexible than the polyester bag. Conversely, a comb made of polyester would be extremely rigid.

There is virtually no end to the polymer combinations that can be devised for a single molecule. Perhaps the most widely exploited is ethylene, already the basis of more than 100 synthetics, and still one of the favorite materials of the polymer chemist.

INFLEXIBLE	UNBRANCHED	UNALIGNED

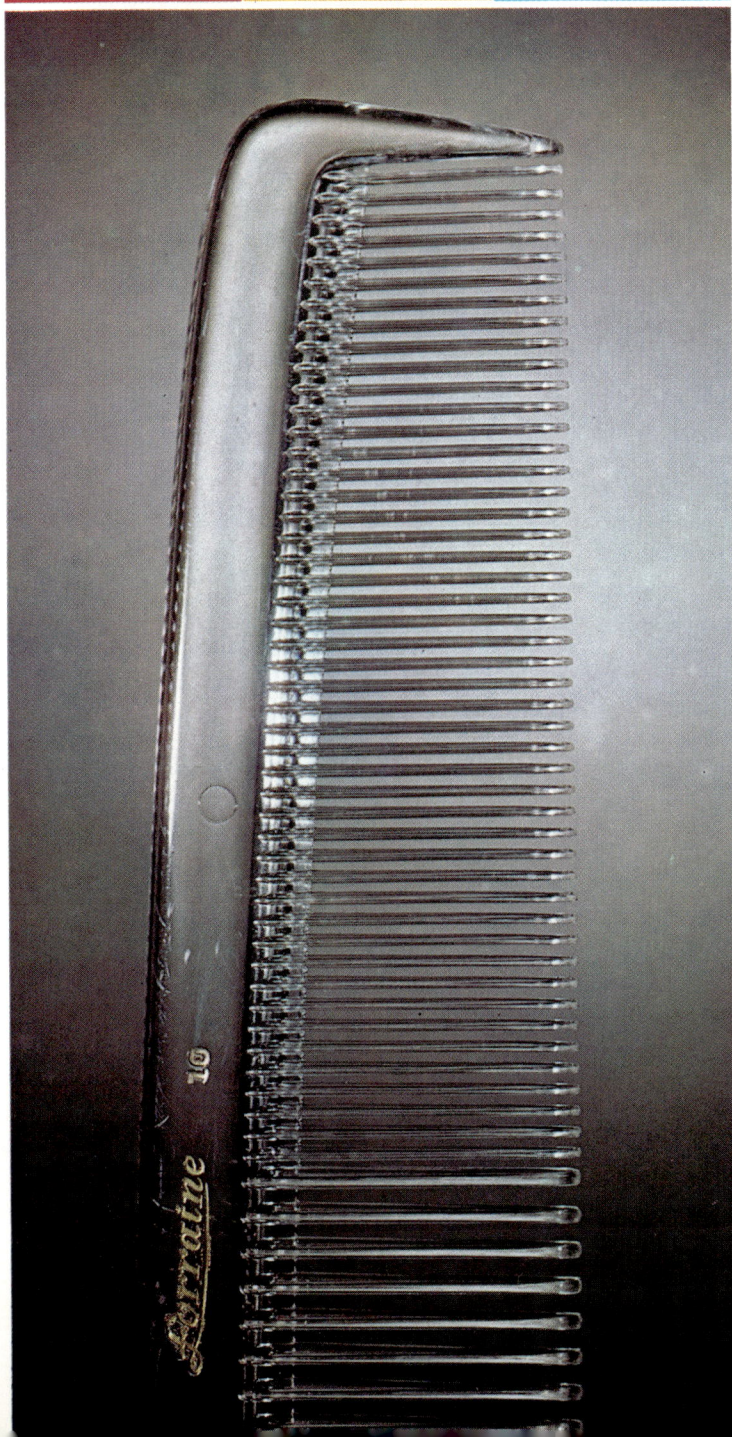

INFLEXIBLE	UNBRANCHED	ALIGNED

4
The
First
Synthetics

A model of Baekeland's laboratory includes the steam-heated still he used to make Bakelite.

ONE DAY IN 1846—so the story goes—Christian Schönbein, a professor of chemistry at Switzerland's University of Basel, was performing some experiments in the kitchen of his home. By accident, he broke a flask in which he had been distilling nitric and sulfuric acids, and the corrosive liquid spilled all over the clean floor. Unable to find a mop, Schönbein wiped up the mess with his wife's cotton apron. Then he washed the apron and hung it in front of the hot stove to dry. But instead of drying, the apron flared up and disappeared. It had been transformed from cotton into guncotton—the base for smokeless powder.

Schönbein's chance discovery was a major factor in a series of extraordinary advances in chemistry that occurred in the second half of the 19th Century. It stimulated the development of a wide range of new materials—not only explosives, but celluloid, rayon, photographic film, cellophane and high-gloss lacquers, to mention but a few. It helped to spark the growth of a gigantic chemical industry in Europe and the United States and, perhaps most important of all, it helped lay the groundwork for the modern science of plastics.

As is often true in science, practical results were achieved before anyone understood the fundamental processes involved. Many years passed before chemists figured out what had happened to Frau Schönbein's apron—that Schönbein had altered the chemical composition of a giant molecule, or polymer, in this case the cellulose molecule which comprises cotton—and thus had changed its properties.

This accidental achievement, momentous in itself, provided the foundation for an even more spectacular triumph 60 years later. In 1907 the Belgian-born American chemist Leo Baekeland combined two kinds of ordinary molecules and broke them down into components, which then reassembled themselves into history's first commercially valuable man-made giant molecule. The new molecule was unique. It had no duplicate in nature. The two substances Baekeland had united produced a third with properties far different from either of the originals. It was not simply a modified version of some pre-existing material, but an entirely new one—the first true synthetic plastic.

Today, chemists can manipulate small molecules almost as if they were building blocks, to produce giant molecules that conform to predetermined specifications; they can thus produce all sorts of miracle products from women's clothing to nose cones of spaceships.

But during the years between Schönbein and Baekeland, the work of experimental chemists was largely empirical. Never understanding fully what they were doing, they compounded, synthesized and analyzed, adding gradually to the store of theoretical knowledge and stacking the shelves with new products. Although they were unable to construct wholly artificial giant molecules, they exploited to the full their new-found skill in altering the chemical composition of several of the giant molecules found in nature.

Of these naturally occurring polymers the one most used was cellulose. The chemists' choice of cellulose is not surprising. It is one of the world's most common raw materials, the substance that makes up the cell walls in all plants and vegetables. It is found in trees and carrots, water lilies and cotton, flax and grass. It provides indigestible but necessary roughage in the human diet, and nourishment for such ruminant animals as sheep and cows. Indirectly, therefore, it also provides nourishment for man: in the process of digestion, sheep and cows convert cellulose into body tissues—meat that turns up on man's dinner table.

But it is not as food that cellulose has its greatest direct value to man. It is tough, fibrous, durable and insoluble in water—qualities which are all related to the size of the molecule of which it is composed. For these reasons, cellulose is valuable for all kinds of uses. The presence of cellulose in wood (comprising roughly 40 to 50 per cent of its solid content) contributes to the rigidity and firmness that make it an excellent building material. The cellulose in jute (65 to 75 per cent) and in hemp (80 to 90 per cent) contribute to their fibrosity and strength, and thus to their value as the raw materials for baskets, mats and rope. The cellulose in flax (80 to 90 per cent) and in cotton (88 to 96 per cent) contribute to their durability, washability and capacity to accept dyes, and thus to their value as woven cloth. And the cellulose in paper—made either from wood pulp or from rags—contributes to its rigidity and durability and to the absorbency that permits it to hold ink.

Plants build cellulose out of one of the most basic of all organic substances—glucose, a simple sugar that is found throughout the vegetable kingdom and in the blood of animals. This relatively small molecule has the chemical formula $C_6H_{12}O_6$; its 24 atoms of carbon, hydrogen and oxygen are arranged in the closed, six-sided structure, called a ring (shown below), that is characteristic of many common organic compounds.

Creating giant molecules

Nature converts glucose into cellulose through an extraordinary process called polymerization—the process which transforms small molecules into giant molecules. To produce one molecule of cellulose, thousands of glucose molecules link themselves together in a long chain, each of them giving up two atoms of hydrogen and one of oxygen—in other words, one molecule of water—in the process. This is an example of the process called condensation polymerization.

Although scientists had synthesized other polymers, including rubber, in all their details, they had not been able to perform the feat with cellulose, principally because there was little understanding of how nature converted glucose into cellulose. In 1964, William Hassid, George A. Barber and Alan D. Elbein of the University of California at Berkeley discovered the secret of this process and duplicated it in the laboratory.

In structure, the cellulose molecule is nothing more than a string of

MOLECULES OF CELLULOSE, the material which comprises the cell walls of all plants, are made of long chains of glucose sugar molecules. Each glucose molecule has a stable, six-atom carbon-and-oxygen ring (brown), with reactive atoms of hydrogen and oxygen projecting off it. By means not fully understood, enzymes within the plant cells cause two glucose molecules to lose, between them, two atoms of hydrogen and one of oxygen (brown box). These atoms are cast off as water, and the glucose rings link up through the single atom of oxygen that remains. The length of the finished molecule varies from plant to plant, but in cotton, for example, the process is repeated over and over until a cellulose molecule some 10,000 glucose units long has been formed.

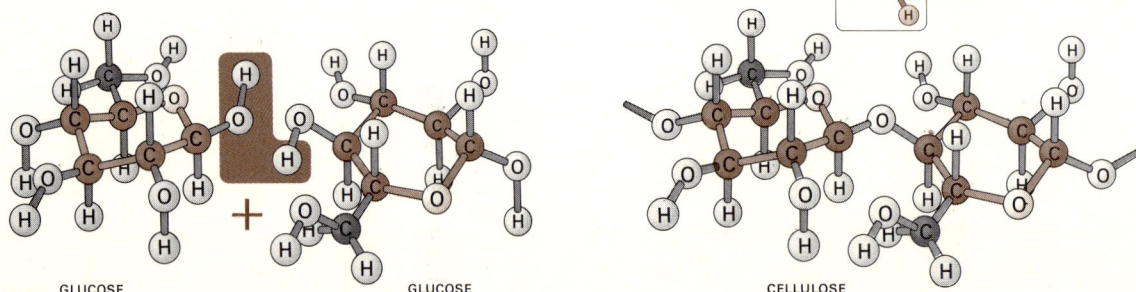

WATER

GLUCOSE + GLUCOSE   CELLULOSE

74

numerous rings, each linked to the next by an atom of oxygen. However, the glucose molecule does not spontaneously combine with others of its kind; therefore, catalysts in the form of plant enzymes are required—both in nature and in the laboratory—to activate the glucose and to create the giant cellulose molecule. As a result of the discovery of suitable enzyme catalysts, man can now create cellulose indistinguishable from that found in nature.

The history-making apron

When Christian Schönbein converted his wife's cotton apron into guncotton with sulfuric and nitric acids, he changed a natural polymer into one that nature never produces. The change was galvanized by the sulfuric acid in the mixture Schönbein used: this chemical did not enter into the structure of the newly created substance, but its presence enabled the cellulose and the nitric acid to combine much more readily than they would otherwise have been able to, and so to form a new giant molecule, cellulose nitrate. In the arrangement of its glucose chain, cellulose nitrate is no different from cellulose, but other atoms have been added to the original chain through the action of the nitric acid. Every glucose ring in the long chain of the cellulose molecule contains three so-called hydroxyl groups—each composed of an atom of oxygen linked to an atom of hydrogen. When cellulose is converted into cellulose nitrate, the hydrogen atoms in these hydroxyl groups disappear, and their places are taken by groups of elements introduced by the nitric acid (HNO_3); these so-called nitro groups, a part of nitric acid, consist of one atom of nitrogen and two of oxygen.

The new substance has very different properties from either cellulose or nitric acid alone. Nitric acid is a corrosive but not inflammable liquid; cellulose, a solid, will burn, but relatively slowly and evenly. However, when the two are combined in cellulose nitrate, the result is an extraordinarily explosive material. The nitro groups in the new molecule are primarily responsible for this effect. Nitrogen is always reluctant to combine with other elements, and the bond between the nitrogen and oxygen atoms in guncotton's nitro groups is, therefore, extremely reactive. Even a tap or a jolt can snap it, and heat causes it to rupture explosively.

Until the discovery of guncotton, the most powerful explosive known was gunpowder, which had been introduced into Europe in the 14th Century, probably from China. Although gunpowder was in wide use, it had one major drawback which no amount of tinkering had been able to remove: it produced so much smoke that contending armies were often unable to see each other in a battle. Guncotton produces much less smoke. In addition, it is three times more powerful than gunpowder.

It is not surprising, therefore, that Christian Schönbein should have

CELLULOSE TRINITRATE (below), derived from natural cellulose (below, left), demonstrates how small changes in the structure of a molecule can radically alter its physical behavior. Cellulose trinitrate has essentially the same structure as cellulose, except that the three hydrogen atoms attached to oxygens (brown), which project from each glucose ring in cellulose, have been replaced in the trinitrate molecule by nitro (O-N-O) groups. This alteration dramatically converts the stable cellulose molecule into guncotton, an explosive three times more powerful than gunpowder.

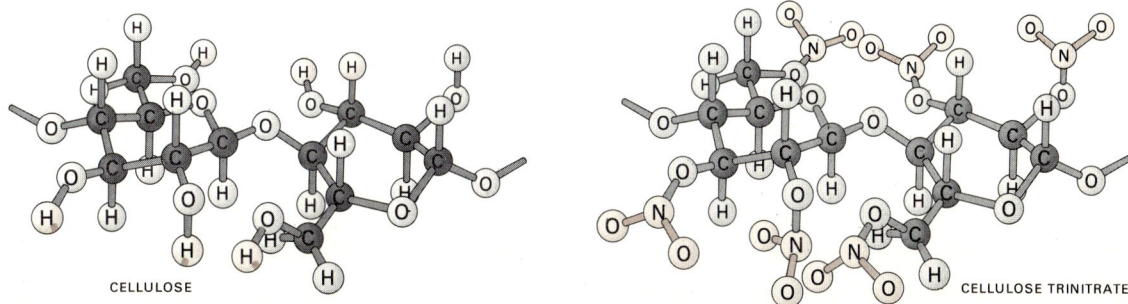

CELLULOSE

CELLULOSE TRINITRATE

been quick to recognize the possibilities of his new material, and quick to make his discovery known. He published his findings in both French and German chemical journals, and gave the formula for the new explosive he had produced. However, as a sound businessman, he did not describe his production methods. Instead, he applied for a patent on his explosive and set out on a search for purchasers. He sold his formula to the Austrian government and to an Englishman, John Taylor. In France, Germany and Russia, there were no takers; instead chemists worked feverishly to duplicate Schönbein's results. Less than a year after publication of his papers, guncotton was being manufactured in all five countries. But in July 1847, only months after Taylor's guncotton plant near London had begun its operations, it was destroyed by an explosion that killed 20 workers. Similar explosions occurred in Russia, Germany and France. By the 1850s, only Austria was still manufacturing guncotton, and even that country stopped in 1862, when both of its plants blew up.

Although these tragedies brought a temporary halt to the commercial manufacture of guncotton, they did not stop scientific investigation of nitrated cellulose. Some scientists, interested in the compound's explosive qualities, continued to look for safer ways of manufacturing guncotton. Others, whose concerns were more scientific, began to investigate the basic characteristics of nitrated cellulose.

Théophile Pelouze, a French scientist and teacher, was in the latter group. Experimenting with different ingredients, he had treated paper with nitric acid and had observed a stunning change of properties. Fascinated with the results, he had encouraged his students to take a closer look at the new material. Late in 1846, Pelouze announced that one of his pupils, Louis Ménard, had made some original observations on nitrated cellulose. The observations, as it turned out, were not only original but of enormous economic importance.

An unrecognized treasure

Ménard discovered the solvent—ether and alcohol—that reduced any nitrated cellulose to a transparent, gelatinous liquid. When exposed to air, this liquid dried to a hard, colorless, transparent film that bore no resemblance to the sheet of paper from which it had been derived. The new substance was called collodion. Until Ménard found the solvent, the only kinds of nitrated cellulose available were those produced by treating such standard materials as cotton, paper and wood directly with nitric acid. Now with the solvent a great array of completely new products was possible.

Ménard never recognized the value of collodion, and the only use to which he could think of putting it turned out to be totally impractical. He was an art enthusiast as well as a chemist, and it occurred to him that collodion might make an effective finish for oil paintings. It turned out that collodion blurs colors and then superimposes its own white on the paint beneath, so this notion had to be abandoned. Ménard lost interest in the substance when he became involved in the revolutionary movement that swept through Europe in 1848, and he never again returned to chemistry.

In time, however, collodion became important in medicine as a dressing for minor wounds and abrasions. When the liquid was applied, it dried to a tough, waterproof finish that kept the skin clean and well protected. The new liquid court plaster gained quick popularity and soon became a standard medicine-chest item.

But collodion's real commercial possibilities remained largely unexplored until the 1860s. Interest was revived by a highly unlikely circumstance—a shortage of elephants. Among the many uses of ivory from the tusks of elephants was the making of billiard balls, and in 1863 the American manufacturing firm of Phelan & Collender offered a $10,000 prize to anyone who could develop a satisfactory substitute for ivory as a billiard-ball material. An Albany, New York, printer, John Wesley Hyatt, began tinkering with a mixture of collodion and camphor, in the hope of winning the prize. He produced a hornlike material that softened on the application of heat, and could be rolled, molded or extruded into any number of different forms. Hyatt called his material "celluloid."

Because of its brittleness, celluloid never proved to be an effective substitute for ivory in billiard balls—a fact which much distressed its inventor, who never abandoned his search for an inexpensive billiard-ball material—but it did have many other uses. Celluloid was used in combs, the handles of brushes and nail files, and for other accouterments the Victorian lady kept on her dressing table. It became material for the Victorian gentleman's collar: it was washable, and far stiffer than any ordinary collar, no matter how heavily starched. Celluloid was used for photographic film, table-tennis balls and a host of other products.

Celluloid was not the only product which came from collodion. When cellulose nitrate was reduced to this liquid form, it became possible to control its rate of burning, and so to tame Schönbein's guncotton. In 1875, Alfred Nobel combined collodion with nitroglycerin to produce blasting gelatin, an explosive much more powerful than the first and more famous explosive he invented, dynamite. (Legend has it that the idea came to him one night when he was kept awake by the throbbing pain of a gash in his finger, which he had covered with collodion.) Nine years later, the French physicist Paul Vieille invented smokeless powder—colloided cellulose nitrate.

A PRACTICAL CHEMIST, Count Hilaire de Chardonnet was the first man to produce a marketable synthetic fiber. The idea of making artificial silk occurred to him when he was a student and became interested in silkworm diseases. After more than 20 years of work, he created "Chardonnet silk"—the forerunner of modern rayon—from cellulose.

"Mother-in-law silk"

Collodion also found more peaceful uses. Because it could be extruded into a fine thread, the idea of weaving it into new fabrics came to Hilaire de Chardonnet, a French nobleman and physiologist who had worked with the great Louis Pasteur in some of his research on the diseases of the silkworm. Chardonnet's new silklike fabric caused a sensation when it was shown publicly for the first time at the Paris Exhibition of 1889. The cloth was as washable as cotton and linen, yet it had all the luster and delicacy of silk. Unfortunately, Chardonnet's artificial silk had one serious drawback which apparently was not noted until he began commercial manufacturing in 1891. Workmen at the textile plants, none of whom seemed to like their wives' mothers very much, called the textile "mother-in-law silk," because it was so inflammable that a spark would send it up in smoke.

Although mother-in-law silk was a failure, two other cellulose fabrics, viscose rayon and acetate rayon, introduced about the same time, were highly successful.

The viscose process was developed in 1892 by the Englishmen Charles F. Cross, Edward J. Bevan and Clayton Beadle. Cross and Bevan were partners in a chemical consulting firm in London, and had done considerable research on manufacturing methods for two major cellulose products—paper and cotton thread. In the course of their work, they investigated the effects of various chemicals on the so-called alkali cellulose, which is produced by reacting cellulose with caustic soda, a form of lye. Among the chemicals with which they experimented was carbon disulfide. Combined with alkali cellulose, it produced a grainy substance, bright orange in color, which could be dissolved in caustic soda and water to produce a syrupy—or viscous—liquid that Cross and Bevan called "viscose." Treatment with acid, it was discovered, converted the viscose back to cellulose again. But it was a cellulose with a difference. The complex viscose process broke the long chain of the giant cellulose molecule at several points, producing two, three or even four molecules from a single one. This regenerated cellulose molecule was still a polymer—a giant molecule. But because it was smaller than the natural cellulose molecule, it was somewhat weaker. Viscose rayon, produced by forcing the syrupy viscose through small holes and into an acid bath, proved to be both softer and safer than mother-in-law silk.

The birth of cellophane

The viscose process also led to the development of another valuable material—cellophane. In 1912, Jacques Brandenberger discovered that viscose could be extruded between slots into an acid bath, in which it hardened into a thin, transparent sheet.

Acetate rayon, a more lustrous fabric than viscose rayon, was developed as a result of work first done in 1865 by the French chemist Paul Schutzenberger. Schutzenberger heated cotton with acetic anhydride (the acid of vinegar, from which the water has been removed), to produce a white powder that could then be dissolved in acetone. Like cellulose nitrate, the cellulose acetate molecule (it is now known) had a different chemical composition from the molecule of natural cellulose although the chain structure was the same. In the acetylating process, as in the nitrating process, the hydroxyl groups in the original

MAKING CELLULOSE TRIACETATE, the material used in rayon and photographic film, involves treating cellulose with acetic acid, the stuff that gives vinegar its sour taste. The acid removes three hydrogen atoms from each glucose ring in the cellulose molecule *(below)* and replaces them with three acetyl groups— $COCH_3$ (the hydrogen atoms and their replacements are shown in brown). The acetyl groups make the molecule strong and flexible.

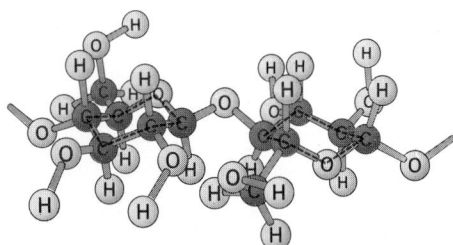

CELLULOSE

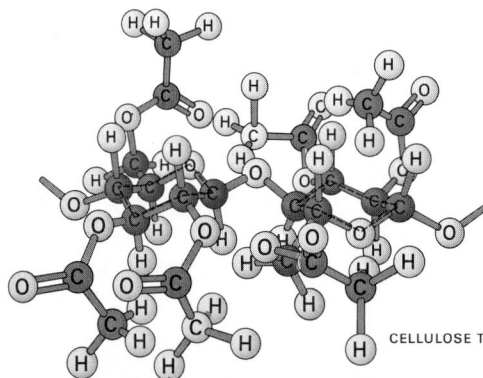

CELLULOSE TRIACETATE

molecule lost their hydrogen atoms: in this case, acetyl groups ($COCH_3$) were substituted for them, making the material nonflammable.

Acetate rayon was put into commercial manufacture in the early years of the 20th Century. But the acetylating process was not widely used until the First World War, when the need arose for a nonflammable dope —or stiffener—which could be used to coat the fabric wings of airplanes. A solution of cellulose acetate proved to be ideal for this purpose. It also formed the basis of photographic films that were a considerable improvement over those made with nitrated cellulose: acetate film was more supple and far less dangerous, since it was not flammable.

Although chemists were able to achieve spectacular changes and variations in the composition of giant molecules during the late 19th and early 20th Centuries, they were still working on an empirical basis, with very little understanding of the actual structure of the molecules with which they were so successfully tampering, and no understanding at all of the processes by which nature produces these polymers. When, in 1907, Leo Baekeland created the first important man-made polymer, he certainly realized the extraordinary nature of his accomplishment. But although he knew the chemical composition of his new substance, and although he knew the methods he had used to produce it, he could not know exactly how the two small molecules with which he began had been converted into a giant molecule.

At the time Baekeland undertook his research, he had already established his reputation as a chemist. While still in his thirties, this talented scientist had invented Velox, the first photographic paper on which pictures could be printed by artificial light. George Eastman, inventor of the Kodak and later president of his own manufacturing company, heard of the new product and made up his mind to buy it. He invited Baekeland to his plant in Rochester to discuss the matter. Baekeland went, having determined, with what he no doubt considered hard-headed business acumen, to ask for $50,000 and to accept no less than $25,000. Eastman offered him a million. And with that fortune in his pocket, Baekeland was able to retire to his private laboratory in Yonkers, New York, where he could pursue his scientific interests unharried by the need to earn a living.

A substitute for insects

Among the projects that interested Baekeland most was the possibility of synthesizing an artificial shellac, which could substitute for the natural product in lacquers, varnishes, waxes and other coatings. Shellac comes from the resinous secretions of *Laccifer lacca*, an insect native to Southeast Asia. It takes about 150,000 of these tiny creatures nearly six months to produce enough resin to yield one pound of shellac. And

NEWBORN INSECT

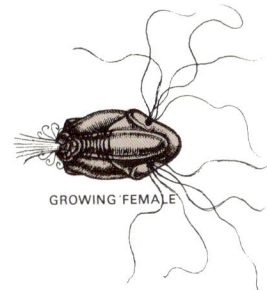

GROWING FEMALE

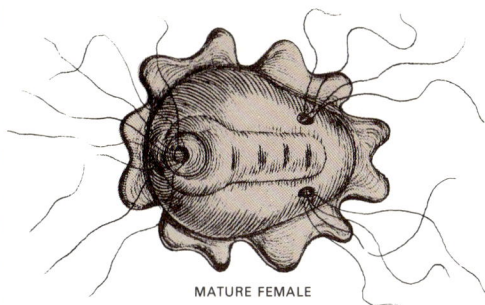

MATURE FEMALE

TINY INDIAN LAC BUGS, magnified here about 20 times, suck sap from trees and secrete it in the form of a resin—one of the few found in nature. The newborn female *(top)* attaches itself to a tree, where it will spend the rest of its life. As it grows *(center)*, it starts secreting a sticky resin derived from the sap, which accumulates until it completely incarcerates the bug *(bottom)*, leaving openings only for a few breathing hairs and for the baby bugs to crawl out. The mature females die after producing a new generation of bugs, and their bodies are collected, ground and processed to make shellac.

the United States alone was importing millions of pounds of shellac each year.

Baekeland's first step was to search through chemical literature to see if it could offer him any guides. He found that in 1871 the famous German chemist Adolf von Baeyer had mixed phenol (carbolic acid, a commonly used disinfectant) with the smelly preservative formaldehyde, to produce a hornlike material. Heated, the mixture foamed, bubbled and stank; cooled, it congealed into a hard, porous and insoluble gray mass—a perfect resin.

To Baeyer and the other 19th Century chemists, these resinous reaction products had been nothing but a plague. In a purely practical sense, they were extremely troublesome: they clogged test tubes and retorts and, since they could not be dissolved, left these receptacles useless. As theoretical problems, they were maddeningly insoluble: the resins could not be analyzed, identified, classified, modified or put to any use. In fact, since all resins are giant molecules, the 19th Century chemists who produced them by accident were anticipating modern science and actually synthesizing polymers in the laboratory. But none of them realized this. In their view, artificially produced resins were to be avoided at all costs; chemical mixtures should be handled gently and carefully, so that they would not give rise to these unpleasant substances.

But Baekeland realized that these resins might be silk purses rather than sows' ears. The very recalcitrance of the phenol-formaldehyde resin might prove its greatest value: the material should be able to withstand weather, wear and chemicals; it might thus form a better coating than the natural lacquers then in use.

Encouraging the undesirable

Baekeland began to search for ways of turning the phenol-formaldehyde resin to use. This meant doing deliberately all the things that earlier chemists had tried to avoid. Instead of attempting to inhibit the reaction which caused the phenol and formaldehyde to combine, Baekeland encouraged it. Instead of cooling the mixture once it was formed, he heated it. To be sure that the higher temperatures would not keep it permanently in the liquid state, he put the mixture under pressure. And finally, he added ingredients to speed the entire process. When the reactions had run their course, he turned over his apparatus to see what he had produced. There, in the vessel, was a clear, hard, amberlike material, molded into the hemispherical shape of the apparatus in which it had been cooked, and faithfully reproducing the pattern of each of the vessel's rivets and seams.

Baekeland called this new material Bakelite. It immediately found —as indeed it still finds—application in virtually every industry. Bakelite

BAKELITE, the first useful synthetic polymer, was once promoted as a lacquer for brass bedsteads, and has since been used for everything from engine parts to billiard balls. It was created in 1907 by Leo Baekeland from phenol and formaldehyde *(below)*. By carefully heating and compressing the mixture, Baekeland got the formaldehyde molecules to lose their oxygen atoms and help provide a carbon link between phenol rings. Once the reaction was started, the molecule grew into a huge network of linked rings.

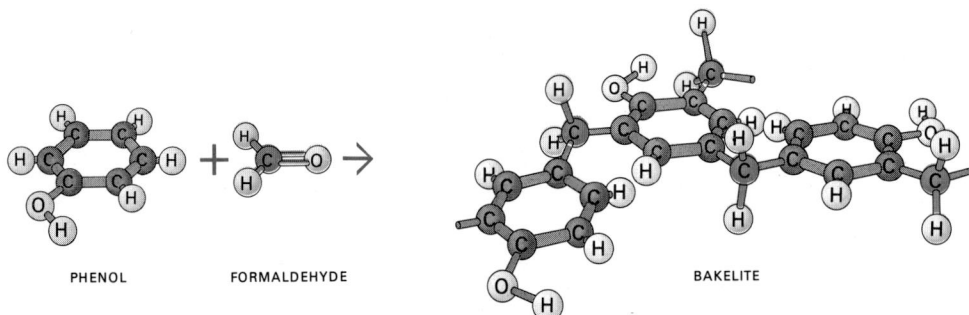

PHENOL   FORMALDEHYDE →   BAKELITE

has a host of desirable properties. It is hard. It can be machined and molded. It is an electrical insulator. It can be colored and dyed. It is enormously strong, yet light in weight. And it is insensitive to heat, acids and weather. It is one of the most versatile materials that man has produced. It is used to make everything from buttons to radio and telephone equipment, from knife handles to coils, switches and distributor heads for automobiles, from counter covering to cameras.

A revolution in chemistry

Bakelite was the first true synthetic, the first giant molecule deliberately created by man. By mixing phenol, which can be derived from coal tar, and formaldehyde, from coal and wood smoke, Baekeland fabricated an entirely new material which cannot be made from any naturally occurring substance. From phenol, whose chemical formula is C_6H_5OH, and from formaldehyde, whose formula is $HCOH$, came the plastic Bakelite with a far different composition: $HOCH_2C_6H_4OH$. Both phenol and formaldehyde are monomers—i.e., simple molecules which build up into polymers—with very simple structures. The first is a version of the common benzene ring, the second the molecule with a T shape often found in hydrocarbons with double bonds. The giant Bakelite molecule, on the other hand, is extremely complex: it consists of a network of benzene rings, each joined to another at one, two or three of its six corners by a chemical group of two atoms of hydrogen and one of carbon.

The scientific leap between Christian Schönbein's manipulation of the cellulose giant molecule and Leo Baekeland's creation of a new and unique giant molecule was truly enormous. Changing the cellulose polymer into cellulose acetate or cellulose nitrate is rather like changing some of the keys on a typewriter so that the machine can be used to type French and Spanish as well as English. The typewriter has been altered, but it is still a typewriter. Turning phenol and formaldehyde into Bakelite is a very different matter. It is like taking a few hairpins and a can opener, melting them down into their components, and reassembling them to form a color television set, complete and in perfect working order. Baekeland took two simple substances and, by combining them, turned them into a complex substance entirely unlike either of the original two and entirely unlike any substance that is found in nature. Cellulose is a natural polymer: the giant molecule is the form into which nature has cast this material and it is this form that gives cellulose many of its useful properties. Bakelite is a man-made polymer, created in the laboratory. When he synthesized this new substance in the first decade of the 20th Century, Leo Baekeland set off a scientific and industrial revolution that is still going on today.

Cellulose: Wood's Wonder-Worker

Cellulose is nature's own giant molecule. It is the fibrous material that every plant from seaweed to the sequoia makes by linking glucose molecules in long chains; the chains are bound together in the fibers that give plants their strength and shape. Men have known for thousands of years how to rearrange cellulose fibers to make paper and linen, but only in the last century did they learn how to dismantle cellulose molecules and rebuild them into tailor-made "natural synthetics"—cellophane, celluloid, film and a growing assortment of other products.

In 1884 the Count de Chardonnet successfully mimicked the silkworm by chemically digesting cellulose and extruding it through holes to form shiny filaments. The result was a type of rayon, the first man-made fabric. Today about seven billion pounds of rayon is produced annually—more than 100 times the world output of silk—and so many new ways are being found to exploit the properties of cellulose that no one dares to predict what may come next. Said Charles F. Cross, a leading experimenter with the substance: "I stand in tongue-tied awe of the cellulose molecule."

CELLULOSE FROM THE FOREST
Spruce logs await shipment by barge down North Carolina's Cape Fear River to a mill where they will be turned into the sheets of pulp that are used in manufacturing rayon, paper and a wide variety of other cellulose products. Most of today's cellulose is extracted from trees; in the United States alone more than 30 million tons of wood is converted into cellulose-rich pulp every year.

The Sources of Cellulose

Every plant is a storehouse of cellulose (an imaginative French chemist once made excellent paper from artichokes), but it can be profitably extracted from only a few plants. Cotton linters, the fibers that cling to cottonseed, were the chief source of commercial cellulose before World War I. But because they are expensive and the supply varies, linters are used today chiefly for high-quality paper.

Wood—especially spruce and southern pine—is now the main cellulose source. It is cheap, abundant and dependable; but since it contains only 40 to 50 per cent cellulose, the substance must be extracted by "pulping." The logs are chipped into flakes and simmered in chemicals that dissolve the tarry lignin, resins and minerals. The remaining pulp, about 93 per cent cellulose, is dried and rolled into sheets—the raw material for paper, rayon and other cellulose products.

A FOREST IN CHIPS

This mountain of wood, reduced from hefty logs to chips roughly the size of breakfast-cereal flakes, will yield more than 20,000 tons of cellulose. The wood was cut from industry-owned forests, which make up about 7 per cent of the half-billion acres of forest in the United States.

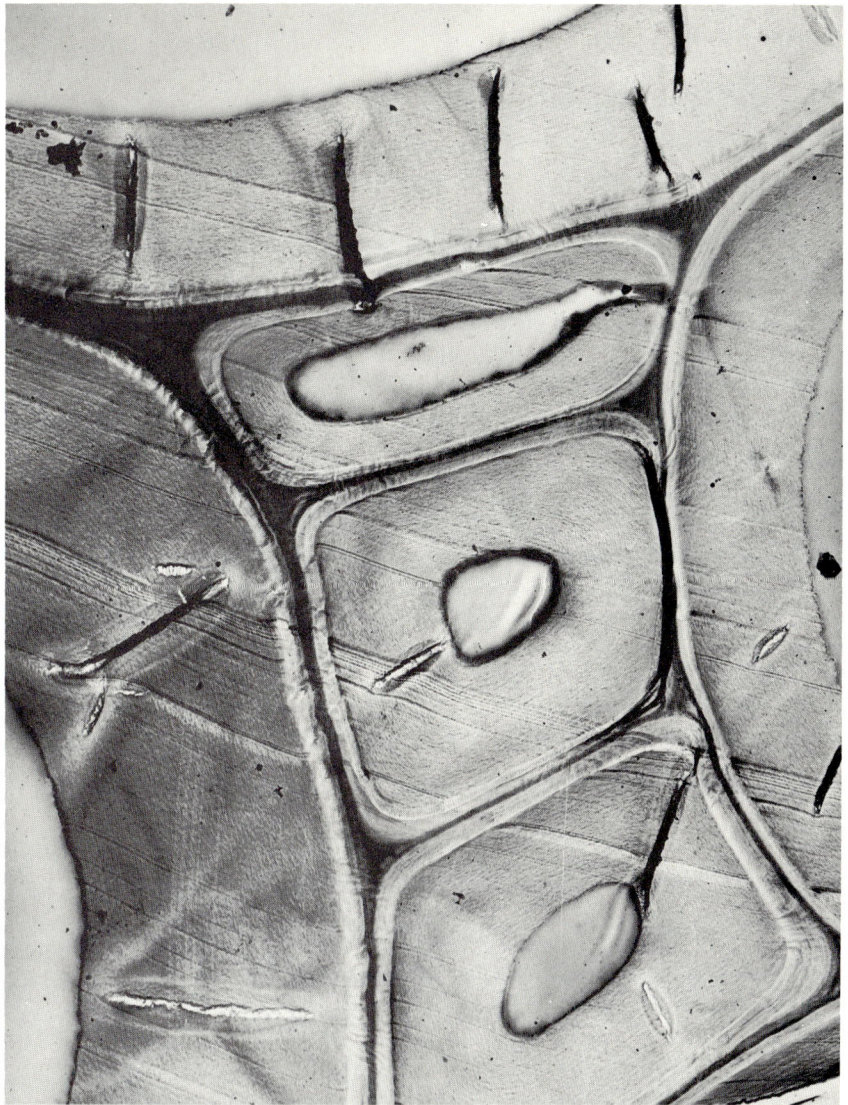

THE TREASURE OF THE WOOD CELLS

A cross section of hemlock chip, magnified 8,000 times, shows the cellulose-rich areas between each cell's wall and its center. The cells, which appear to be compressed into odd forms, are actually shaped like elongated bananas and are staggered throughout the wood; each cell is being seen in cross section at a different point.

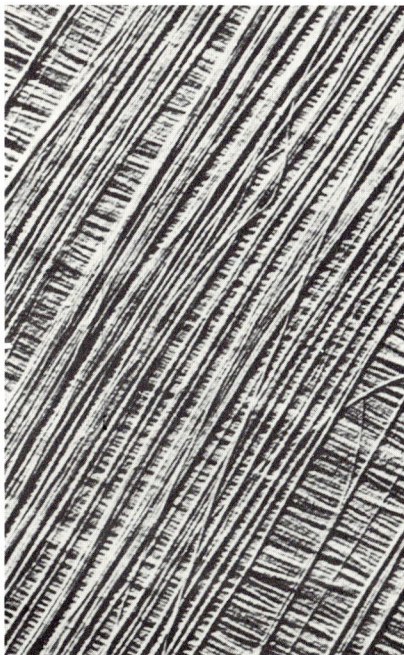

FIBERS FROM THE SEA

When the microfibrils (subunits of cellulose fibers) of seaweed are magnified 18,000 times, they show a rare crosshatched arrangement— in most plants they lie in roughly the same direction. Seaweed is potentially a good source of cellulose, and may become more widely used if wood can no longer meet industrial demand.

The Many Faces of Nature's Polymer

Cellulose products are divided into two families, one ancient and the other relatively new. In the old group are cotton cloth, linen and paper, all products in which the cellulose molecules remain intact. Cotton cloth was used for shrouds in pre-Inca Peru. Ancient Egyptians flayed flax into long fibers which they wove into linen. The Chinese boiled mulberry bark in lye and dried the gummy liquid on screens to make paper. Medieval craftsmen made paper from rags (below); the same material is still used for high-quality stock. With the invention of the printing press, a demand arose for cheap paper, and this led to processes utilizing straw, seaweed and cornhusks. Today wood is the cheapest and most widely available source of cellulose for paper.

The newer cellulose products—celluloid, cellophane and rayon—were made possible by the chemical revolution of the 19th Century, in which man learned to alter cellulose molecules for his own purposes. Many products were developed in Europe, but they reached their peak of success in the U.S. Cellophane, for instance, was so utilitarian that by 1927 it had been used to wrap everything from peanuts to the original of the Declaration of Independence.

AN EARLY PAPERMAKER
A medieval craftsman soaks cellulose pulp while an apprentice carries a stack of finished paper to a storeroom. The pulp was made from rags boiled in a mixture of water and ashes; wood pulp was not used in Europe until 1840.

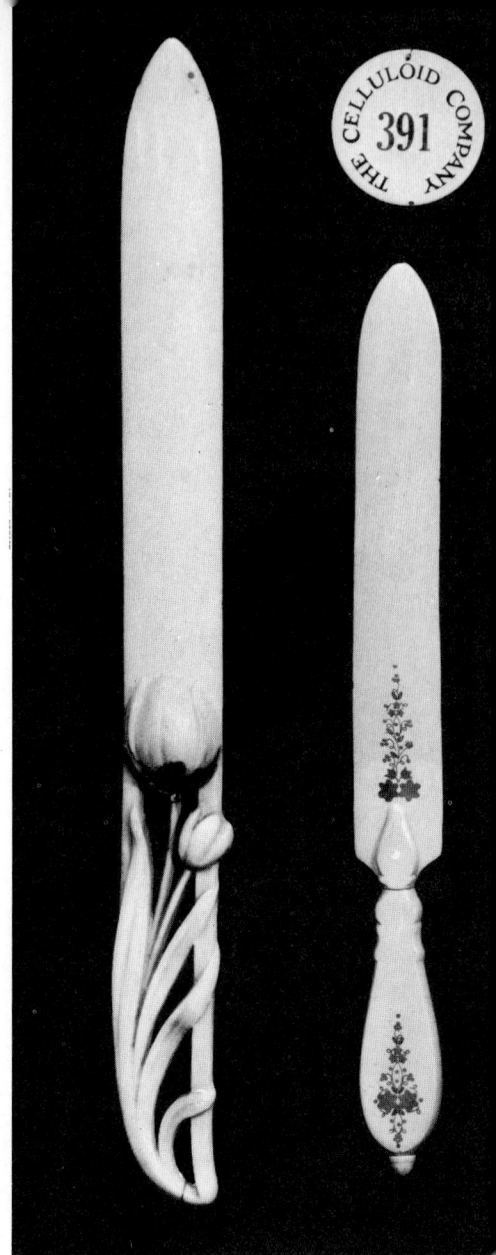

WHEN CELLULOID WAS KING
Celluloid, an inflammable amalgam of cellulose, nitric acid and camphor, was the first modern plastic. Invented in 1868 by John Wesley Hyatt, by the end of the century it was being molded into dice, dominoes, "rubber" stamps (center, top) and even medals for Coxey's Army (center, bottom). But consumers were sometimes seared by blazing celluloid collars and eyeglass frames; by the 1950s it had been largely displaced by cheaper, safer petroleum-based plastics.

MILES AND MILES OF CELLOPHANE
Giant rolls of cellophane are inspected at a Kansas plant that produces enough in a year to circle the earth 40 times with a 15-inch-wide strip. Cellophane is chemically identical with viscose rayon, but instead of being extruded through holes it is squeezed between rollers.

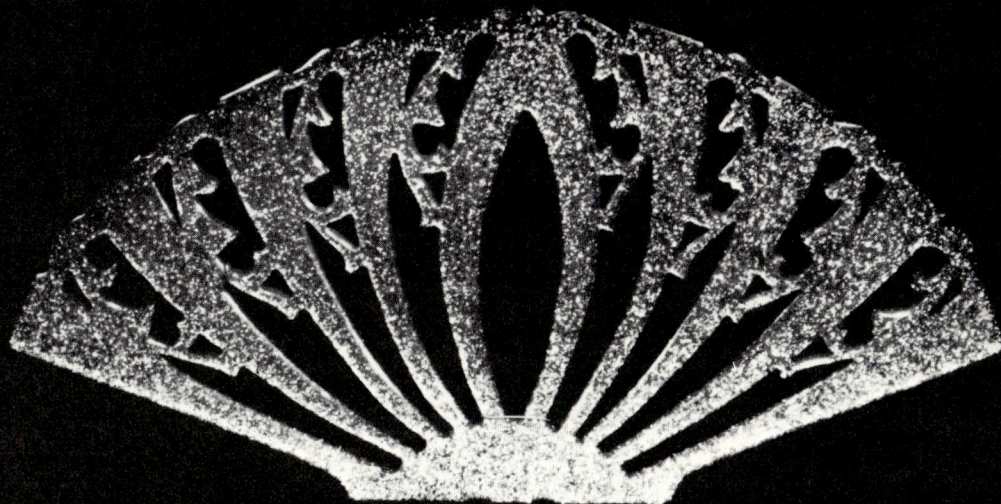

YOU

CELLULOID

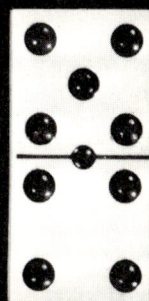

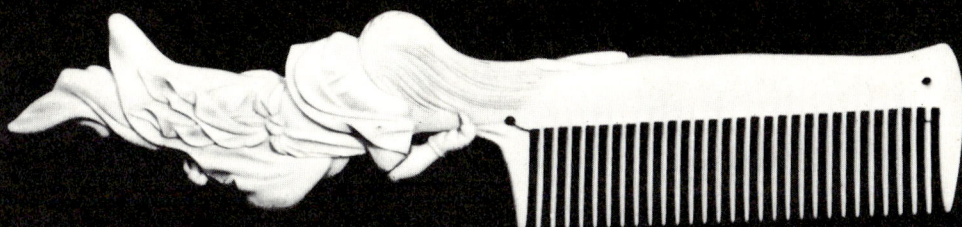

VISCOSE PROCESS

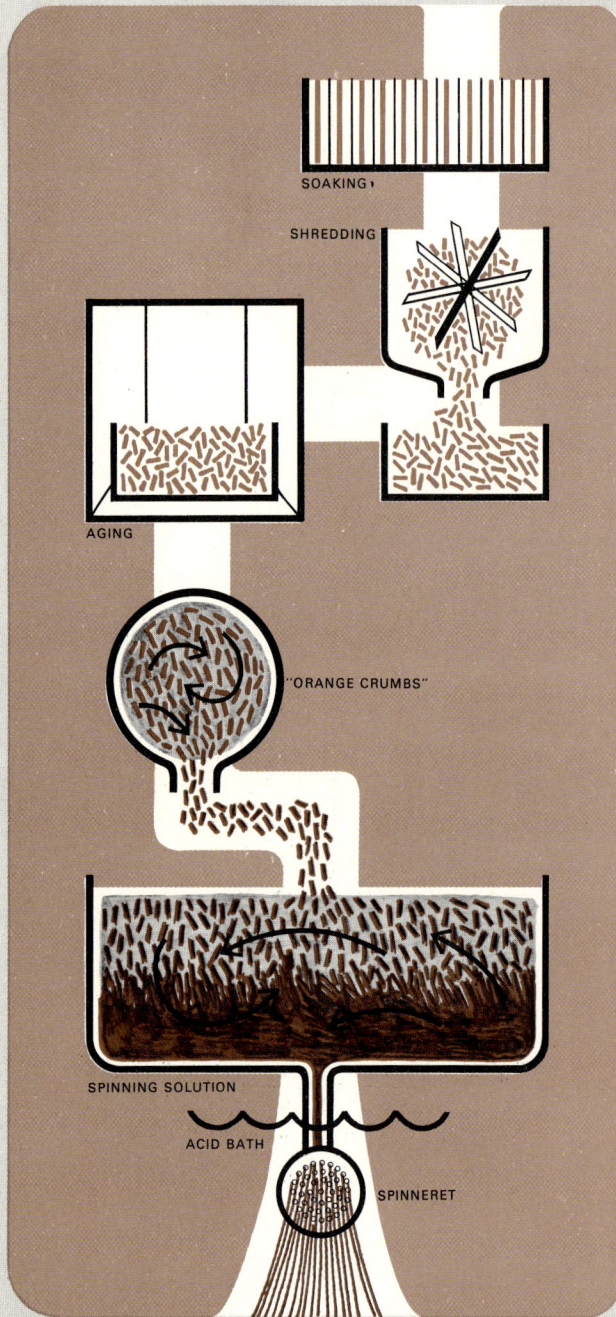

SOAKING

SHREDDING

AGING

"ORANGE CRUMBS"

SPINNING SOLUTION

ACID BATH

SPINNERET

ACETATE PROCESS

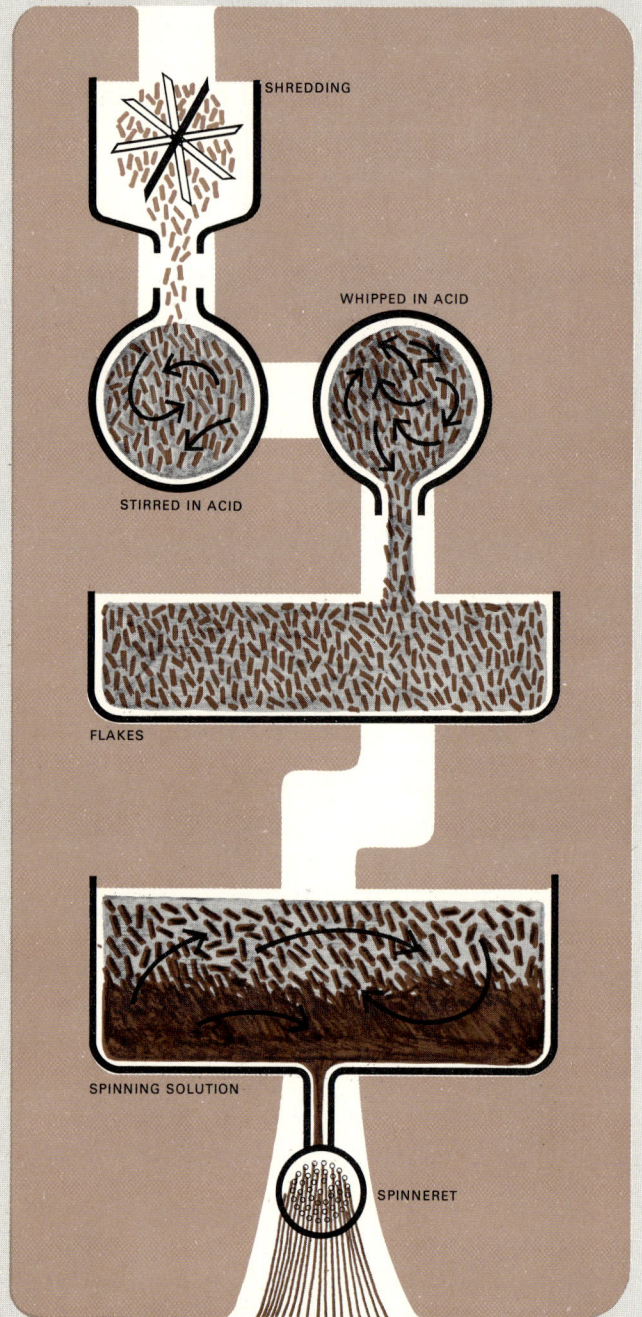

SHREDDING

WHIPPED IN ACID

STIRRED IN ACID

FLAKES

SPINNING SOLUTION

SPINNERET

FROM PULP TO VISCOSE
Viscose rayon begins as pure cellulose and ends
as pure cellulose—but only after it has under-
gone several physical and chemical changes.
The pulp is first soaked in lye, which transforms
it into alkali cellulose; it is then mixed with
carbon disulfide and turned into soluble "orange
crumbs." Finally, it is regenerated into pure cellulose, extruded as a spray of solid rayon filaments while it passes through sulfuric acid.

FROM PULP TO ACETATE
Acetate rayon starts with the same natural
cellulose used in the viscose process, but the
finished product is cellulose acetate—every molecule has been chemically changed by acetic
acid. The powdery white flakes of cellulose acetate are dissolved in acetone, and the solution
is extruded through a sprinklerlike spinneret
directly into the air. As the acetone evaporates, fine filaments of acetate rayon are formed.

Two Roads to Rayon-making

For 48 million years silkworms have digested the cellulose in mulberry leaves and extruded fine filaments of silk—for thousands of years the most luxurious fabric known to man.

In 1664 a British physicist named Robert Hooke prophesied rich rewards to the person who could imitate "this dumb little worm." But man knew of no way to make thread of cellulose until the discovery of nitrocellulose—guncotton—in 1846. In 1884 the Count de Chardonnet, by dissolving nitrocellulose and forcing it through tiny holes, made a lustrous "artificial silk"—the first rayon. But after a young lady's artificial-silk dress disappeared in a puff of smoke, nitrocellulose rayon was supplanted by two relatively fireproof varieties—viscose rayon, in which the finished product is pure cellulose, and acetate rayon, in which the cellulose is altered by acetic acid. On the following pages methods of manufacturing both types of rayon are examined in detail. Photographs of the cellulose fibers, magnified 500 times—the first such pictures ever taken during processing—show their structure at every step.

VISCOSE PROCESS

PULP BY THE BUNDLE . . .

Cellulose pulp sheets are rolled to the scales in the Asheville, North Carolina, plant of the American Enka Corp. The newspaper-sized sheets feel like cardboard and are about a tenth of an inch thick. This plant uses more than 200 million pounds of pulp a year to make cord for an estimated 75 million automobile tires plus about 50 million pounds of rayon.

ACETATE PROCESS

. . . AND BY THE ROLL

Giant rolls of sheet pulp are fed into a cutting machine in a factory in Rock Hill, South Carolina, a plant of the Celanese Corporation, the largest producer of acetate rayon in the world. About 565 million pounds of acetate is produced in the United States each year, most of it destined for the clothing industry. It is widely used for women's dainty apparel.

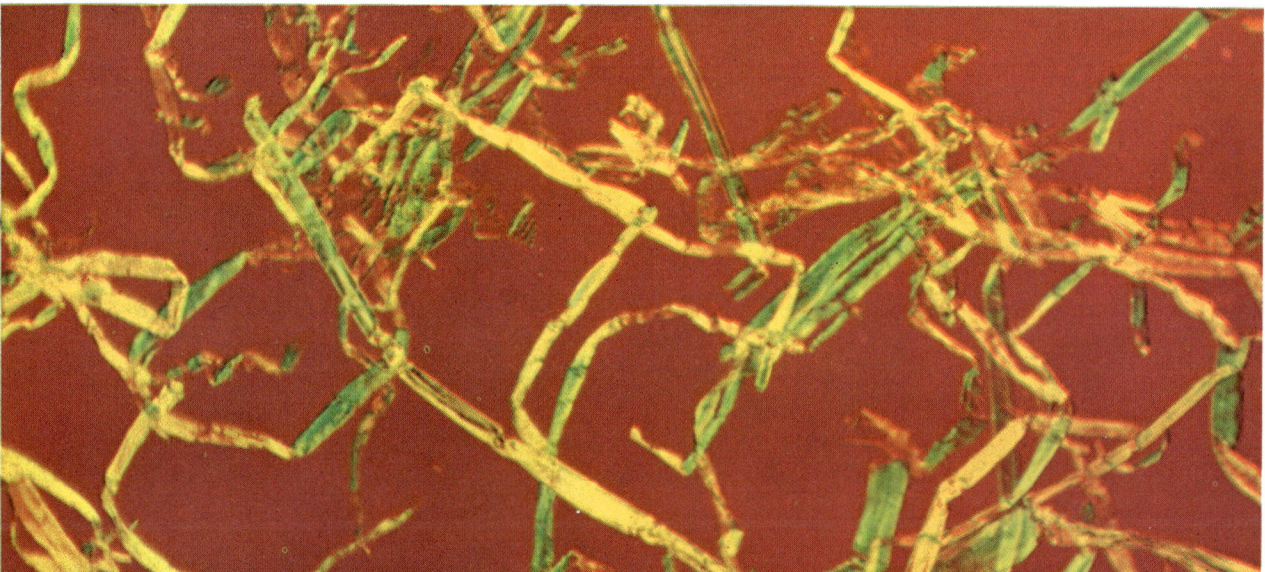

The once-symmetrical cellulose fibers are churned into chaotic disarray by the pulping process, but they are still sturdy and intact.

Stubborn Bundles of "Spaghetti"

The very properties that make cellulose so valuable—its strength, chemical inertness and insolubility—make it extraordinarily difficult to break down into a uniformly syrupy solution that can be forced through the holes of a spinneret to form rayon filaments.

A single cellulose fiber might be compared to so many stiff sticks of spaghetti, firmly stuck together before cooking. Each stick represents one cellulose molecule, and the glue holding them together is hydrogen bonding, the powerful link that locks

VISCOSE PROCESS

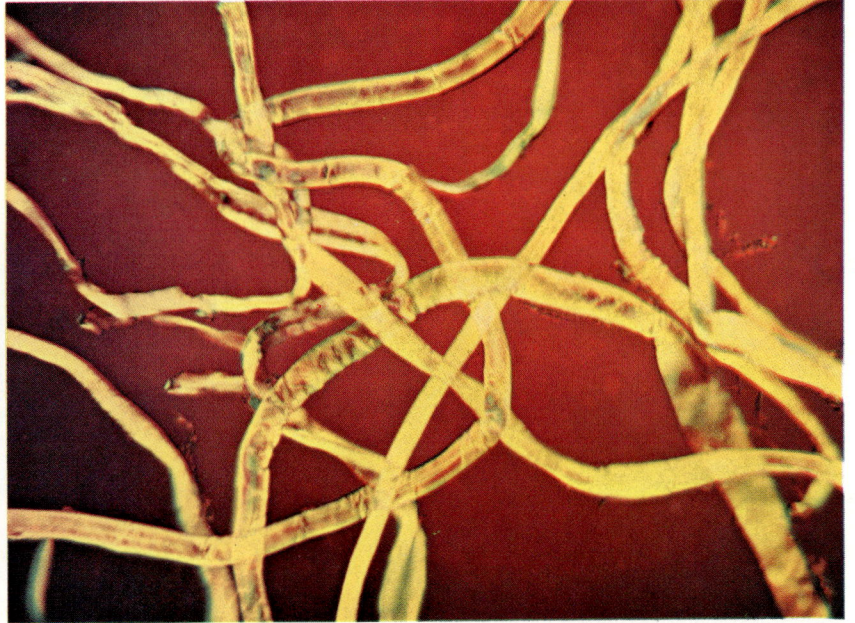

AN IMMERSION IN LYE
In the first step of turning cellulose into fabric, sheets of pulp are placed in each of 30 compartments of a steeping press. After the pulp has been well saturated by a strong lye solution for about two hours, the press contracts like a steel accordion and wrings out most of the caustic solution. At this point the cellulose fibers *(right)* appear swollen because they have begun to separate; otherwise, however, they are still intact.

AN AVALANCHE OF "WHITE CRUMBS"
Bits of pulp, called "white crumbs" by viscose workers but actually about the consistency of cottage cheese, are raked into a cart as they tumble from a shredding machine at the rate of about 900 pounds a minute. Powerful rotary blades have chewed the pulp into soft, fluffy particles. The gaps between the cellulose fibers *(right)* have now widened, allowing the remaining lye solution to seep deeply into every fiber.

the oxygen atom in one cellulose molecule to its neighbor's hydrogen. The first step in breaking down cellulose fibers, therefore, is to unlock the molecules. Just as the housewife throws the stubborn spaghetti into a pot of water, so the cellulose molecules are soaked apart. In the viscose process, the pulp is immersed in lye, or sodium hydroxide; the hydrogen bonds are broken as the oxygen atoms release their neighbors' hydrogen atoms and seize on the more attractive sodium atoms. In the acetate process, the oxygen atoms are lured away from the hydrogen atoms by molecules of acetic acid. In each case the result is the same: cellulose molecules sag apart, their fibers soft and swollen, and the cellulose is ready for its next transformation on the twin roads to rayon.

ACETATE PROCESS

A BATH IN ACID
Flakes of shredded pulp are mixed with acetic acid—the liquid that gives vinegar its sour taste and smell—and slowly churned into a mash in one of 30 closet-sized tanks *(right)*. The unmasked worker at left has become accustomed to the pungent acid fumes through continued exposure, but for anyone else they are utterly intolerable: the photographs at upper right and lower left had to be taken by remote control.

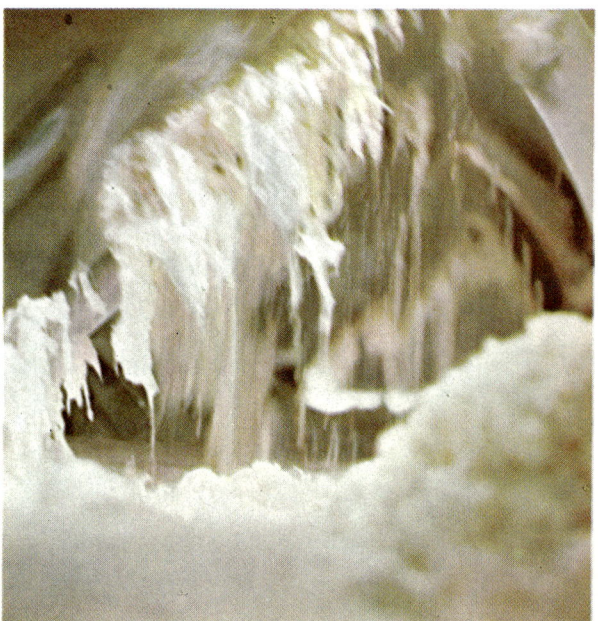

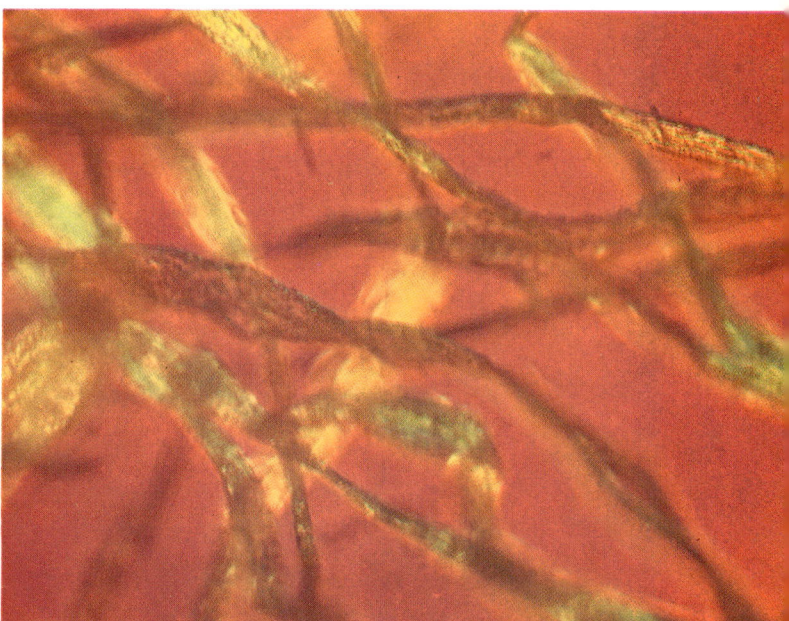

A VIOLENT WHIPPING
The pulp-and-acid mixture is piped to a compartment where it is beaten into a frothy foam *(left)* by high-speed agitators resembling washing-machine blades. At this point the fibers are weakening rapidly as the acid ruptures the hydrogen bonds between the cellulose molecules. The mixture is now cellulose acetate; it will go through several additional steps, but its chemical composition will not be changed.

Cutting Molecules Down to Size

Each cellulose molecule in wood pulp contains about 1,800 glucose units—making it far too long to be spun into thread. So the molecules must be shortened to about 400 units before being processed as rayon.

In the original process of linking together to form cellulose, each unit of glucose gave off two atoms of hydrogen and one of oxygen—one molecule of water (H_2O). To break the bond, the process is reversed: two atoms of hydrogen and one of oxygen are reintroduced to each glucose unit. The first step is accomplished in the earlier process of "unsticking" the cellulose molecules: the hydrogen atoms become available when the hydrogen bonds that hold the fiber together are broken. Next oxygen is supplied, simply by exposing the pulp to the air. The process must be carefully controlled: if the molecules get too short they will form weak thread.

VISCOSE PROCESS

WHERE MOLECULES ARE AGED
Carts filled with white crumbs fill a room the size of a football field. After about two days of aging, each of the original cellulose molecules has broken into four or five shorter molecules.

THE BROKEN FIBERS
The aged and broken cellulose fibers, once as resilient as willow switches, are now as limp as wilted lettuce. Sapped of their elasticity, the fibers compress under their own weight.

A HANDFUL OF FLAKES
Flakes of cellulose acetate, as dry as soap powder, flutter from a dryer where blasts of hot air evaporated the excess acetic acid and added oxygen to begin the molecule breakdown.

AFTER THE ACID TEST
Amorphous blobs of color are all that remain of the cellulose acetate molecules after the cellulose fibers have been broken by the acetic acid. Each of the clusters at left is a single flake.

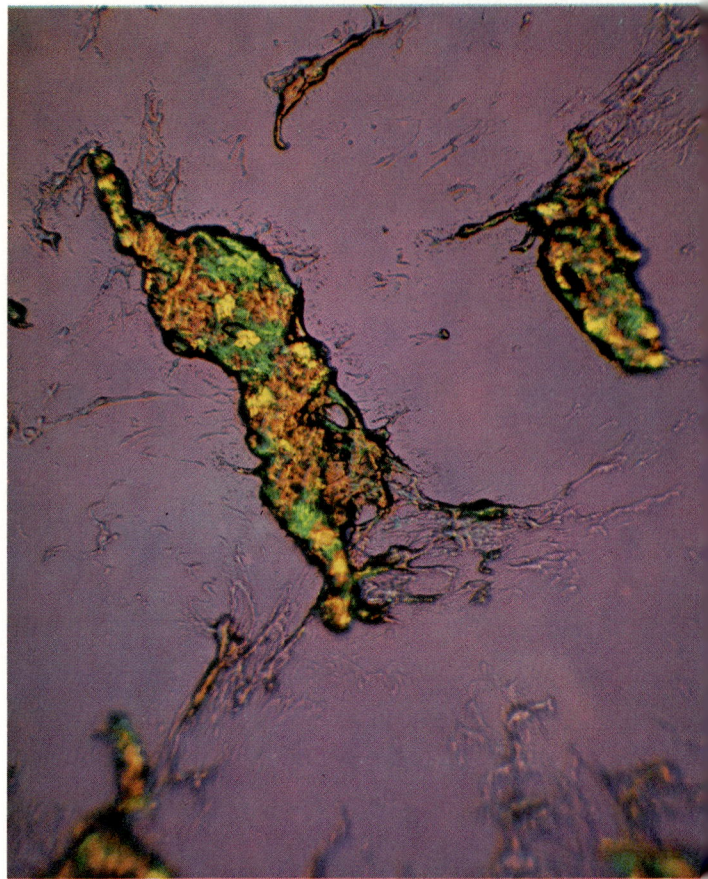

A TORRENT OF ORANGE

Colorful crumbs, transformed into alkali cellulose xanthate (Greek for "yellow") after being mixed with carbon disulfide for two hours, tumble from a huge churn directly into a dissolving tank *(below)*. The fibrous structure of the cellulose has virtually disappeared, as the photomicrograph at right indicates, and the crumbs are now readily soluble in water.

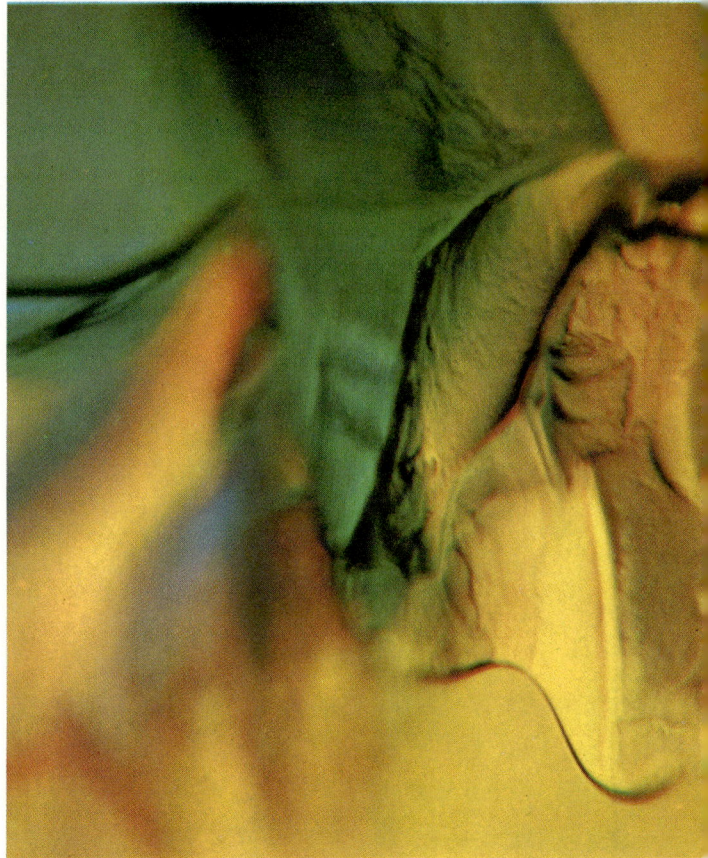

A THICKENING SOLUTION

In great vats containing a dilute lye solution, the orange crumbs dissolve and the cellulose molecules disperse evenly throughout the liquid *(right)*. During the next two or three days, the thick solution ages sufficiently so that it can be spun into threads. But the rate of progress varies from batch to batch; accordingly, tests are run every hour *(above)*. "Making viscose rayon," one expert has said, "is more of an art than a science."

Making Jelly
from Cellulose

Once the cellulose fibers have been broken into separate molecules, and the molecules have been reduced to the proper sizes, one important step remains in both processes before the cellulose can be spun into thread. The crumbs of viscose and flakes of acetate are still solids; they must be changed into liquids.

At this point, the white crumbs in the viscose process are alkali cellulose, which is insoluble in water. But when they are mixed with carbon disulfide they become water-soluble orange crumbs. If they were dissolved in water alone, however, the solution would still be unsuitable for spinning. Instead, they are dissolved in a weak lye solution—the very same chemical that formed the alkali cellulose in the first place. This time the lye gradually turns the crumbs into a gluey viscose, ready to be spun into threads when it reaches the approximate consistency of honey.

In the acetate process, the flakes of cellulose acetate are dissolved in acetone, a powerful industrial solvent that is sometimes used as nail-polish remover. The resulting acetone solution is just about as thick as the viscose solution.

Before being extruded and spun into thread, both solutions are strained to remove impurities, and are subjected to a vacuum to remove bubbles.

A SPIN IN ACETONE

Whirling in an acetone mixer *(above),* the cellulose acetate flakes are whipped until they dissolve into a thick, spinning solution. (Because acetone is both explosive and inflammable, the mixing process is ordinarily conducted in strong, fireproof tanks; this picture is a special demonstration.) Once mixed, the solution is kept in airtight containers until extruded. If it should accidentally be exposed to air for even an instant, the acetone would start to evaporate at once, creating particles of solid cellulose acetate and kaleidoscopic bursts of color *(below).*

Spinning a Colorful Yarn

Creating rayon filaments from a spinning cellulose solution is like turning on a shower and getting streams of ice instead of water. In both the viscose and acetate processes, the thick solutions are forced through the tiny holes of spinnerets, which closely resemble shower heads. A spinneret may extrude as many as 50 filaments. All of these are stretched and twisted into a single thread: a thread of many thin filaments is stronger and more flexible than a single thick filament.

In the viscose process, the spinnerets face upward, immersed in tanks of sulfuric acid *(above, left)*. The acid converts the solution's lye and carbon disulfide into salts; filaments of pure cellulose emerge from the bath and are wound on overhead spools *(above, right)*. In the acetate process, the spinning solution is propelled downward, directly into the air. The volatile acetone evaporates almost immediately, leaving strands of cellulose acetate to be reeled onto a bank of spindles located several yards below the spinnerets *(below, right)*.

Dye may be added to either spinning solution, thus becoming an integral part of the spun thread *(opposite)*. This gives both viscose and acetate remarkable resistance to fading. Because acetate accepts some dyes and rejects others, it offers some intriguing possibilities. For example, white acetate thread can be woven with cotton—which it closely resembles—but when the entire fabric is dyed only the cotton is affected, leaving an attractive pinstripe pattern.

VISCOSE PROCESS

UPWARD THROUGH ACID
Viscose filaments, emerging in a sulfuric-acid bath, merge into a single thread as they are pulled over a guide spindle. The thread *(opposite, above)* is wound on spools and washed and dried before being woven into fabric or chopped into a fluffy material for blending with other fabrics and for filling cushions. Viscose rayon's durability makes it ideal for tire cord. It is also widely used for long-wearing men's clothing after being treated to resist wrinkling and shrinking.

ACETATE PROCESS

DOWNWARD THROUGH AIR
Acetate filaments, dropping from a spinneret, become solid as the acetone in the solution evaporates. For the first few feet the filaments fall through cylinders where hot air evaporates the acetone. Acetate rayon *(opposite, below)* holds a permanent crease, dries quickly and resists wrinkling, shrinking and mildew, all of which makes it excellent for clothing fabrics.

5
"From Coal, Air and Water"

A sports balloon, one of numerous new products made of nylon, soars over a California desert.

WHEN LEO BAEKELAND pried Bakelite out of his reaction vessel, he was unknowingly extracting the future from the past. Behind him lay a long history of empirical accomplishments; ahead of him was a new science. From the knowledge of atomic structure that would be developed in a single generation were to come custom-tailored giant molecules, each different from the next, each unique in nature, many of them useful in ways that were not dreamed of by their creators.

A combination of experimental discovery and spectacular mathematical analysis made the first quarter of the 20th Century one of the most exciting periods in the history of science. Through one of the most critical of the experiments, chemists achieved a technique for "seeing" the structures of the molecules they had been working with—thus eliminating guesswork about the arrangement of atoms in molecules. Even more significant was the analytical power of quantum mechanics, the mathematical tool that revealed why atoms gave a particular structure to a given kind of molecule. The theory foretold not only what compounds could be made but also the difficulties that would probably be encountered in synthesis, and the qualities of the finished product. It was the dawn of a new age in which man for the first time assumed control over some of the fundamental materials of nature.

The major objective of chemists of the 18th and much of the 19th Centuries was well defined by Antoine Lavoisier. "Chemistry," he wrote, "marches toward its goal by dividing, subdividing and subdividing still again." Scientists were primarily concerned with the basic problem of analysis. They sought answers to two questions: first, "What does it contain?" and second, "How much of each element is contained?"

In Lavoisier's day these questions were sufficiently challenging to occupy a man's efforts for a lifetime. Slowly, and with many hesitations and backtrackings, the chemists puzzled out the differences between an element and a compound; by the turn of the 19th Century, thanks to John Dalton, they had arrived at a primitive kind of atomic theory. Dalton's hard, tiny spheres of fixed weight were a far cry from the vibrating, indefinite quantum-mechanical atoms of modern physics, but his atomic theory enabled him to calculate—although sometimes inaccurately—the relative weights of various atoms.

With Dalton's table of atomic weights—plus the painstaking analyses of generations of chemists who followed—it was possible by the 19th Century to deal with a large number of chemical compounds. A chemist could now take apart and put together again almost all of the simpler molecules that make up the environment of life: the soils and minerals of the earth, the waters of rivers and oceans, and the gases of the atmosphere. But when it came to the giant molecules of life itself, chemical analysis was indeterminate. Compounds refused to fall into neat formulas and, worst of all, they tended to change and flow into each other in ways that could not be duplicated in the glassware of the laboratory.

Despite some notable analyses of organic compounds, chemists of the 19th Century lacked one vital prerequisite—an understanding of the true chemical design of the giant molecule. With the theory of valence and Kekulé's theory of chain and ring structures, the chemists of the day were groping toward an understanding of the ways molecules were put together. Even at the beginning of the present century chemistry, and organic chemistry in particular, was still an empirical science, a science in which experiments were largely based on the stubborn discipline of trial and error.

In chemistry more than in most scientific fields, there is often a great lag between the discovery of the way something happens and the explanation of why it does. Just as any good cook can bake bread without understanding the chemical reactions which take place, and can even provide simple directions that will enable someone else to duplicate the accomplishment, so Leo Baekeland, trained chemist though he was, had the recipe for making a giant molecule, but no clear concept of the processes involved.

But even as Baekeland was perfecting his recipes, adjusting the temperatures and pressures in his cooking pots and varying his proportions of phenol and formaldehyde, a chemical revolution was in the making. In half a dozen countries a new breed of scientist was striking directly to the heart of the chemist's problem: determining the actual structures of the molecules and their component atoms.

Photographing a molecule

At the turn of the century, chemists were still guessing—with some success—at the shapes of molecules. None of them had ever seen a molecule, of course, but they pieced together bits of information found in the test-tube reactions. Then in 1912 Max von Laue, a professor of physics at the University of Munich, used X-rays to shine a spotlight into the heart of a molecule and to photograph its structure. He and two assistants placed a slice of zinc ore in front of a photographic plate and then subjected the sample to a beam of X-rays. When the plate was developed, a pattern of spots appeared. Von Laue tried the same experiment with a copper sulfate crystal and obtained a different pattern. More experiments showed him that various crystals, submitted to X-rays, produced different patterns—but the same crystals always produced the same design.

Von Laue had finally found a way to decipher the true arrangement of atoms within a molecule. When X-rays hit the electronic clouds of atoms, the rays were diffracted—scattered—and the directions were determined by the positions of the atoms inside the molecules. The effect might be compared in a very rough way to what happens when a beam of visible light strikes the prisms of a crystal chandelier: as the light bounces off the separate prisms, they throw light in all directions, making spots on the wall and floor and ceiling. The pattern of these spots is determined by the arrangement of prisms on the chandelier. A patient and skillful observer could, by measuring the light spots and applying the formulas of optics, learn the number and arrangement of prisms—and thus the overall structure of the chandelier—even if he had never seen the chandelier itself. Using a similar approach and considering the depth to which X-rays penetrated the crystals, von Laue studied the patterns of spots made on the film and calculated the arrangement of atoms inside the molecule of each sample. Much of the mystery of molecular structure had been solved.

Meanwhile, experimental and theoretical physicists, such as Lord Ernest Rutherford, Erwin Schrödinger, Niels Bohr and Werner Heisenberg, were using a variety of techniques and mathematical analyses to study structure in the still-smaller world of the atom. In less than a quarter of a century—roughly from 1905 to 1929—the broad systematic outlines of atomic structure were established, not in terms of theory and half-philosophical concept, but in the highly concrete terms of mass and energy, electrical charge and electromagnetic wavelength, terms which could be firmly nailed down by measurement.

The atom's electron shells

The new concept of the atom, upon which molecular structure depends, is mathematical rather than material. A useful drawing can be made of an atom as long as it is remembered that the drawing is not a literal diagram but merely an attempt to explain a mathematical idea in visual terms. Such a picture shows an atom with a dense, positively charged nucleus surrounded by a number of negatively charged electrons. The electrons can be thought of as existing on energy levels, each like a vaguely defined shell. Depending upon the kind of atom, it may have a single shell or as many as seven, each fitting inside one other like the layers of an onion.

The atom is held together by electrical force—by the attraction between the positive charges, concentrated in the nucleus, and the negative charges of the surrounding electrons. This same force provides the major bonds which bind atoms into the composite of the molecule. But since a molecule has many atoms, it has several sources of positive charge—the nuclei of the various atoms—and these interact with the negative charge created by the several clouds of atomic electrons. The connection is formed by the electrons that provide the valence of each atom, i.e., the "arms" that enable several atoms to grasp one another and thus create a molecule. The electrons from the outer shells—which are usually the only electrons that serve as molecular fasteners—may be shared by the atoms in a molecule or traded back and forth between

SYMMETRICAL PATTERNS like those below, made by directing X-ray beams through crystals of silicon, tungsten and topaz onto photographic film, are used by physicists to study the arrangement of atoms in crystals. The highly accurate technique, called X-ray diffraction, produces characteristic patterns of spots for each crystal. By noting symmetrical segments, indicated by lines, and measuring the distances between the dots, researchers can calculate the structure of the crystal.

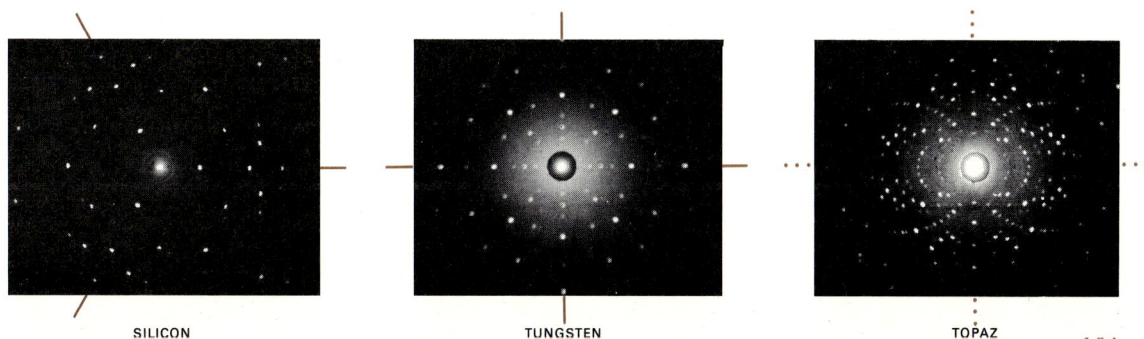

SILICON

TUNGSTEN

TOPAZ

them to form the chemical bonds that hold the molecule together.

The variety of bonds and chemical constituents explains why one substance may differ from another in hardness, tensile strength, conductivity, solubility and other physical properties.

For example, an ionic bond—an attachment between differently charged atoms—accounts for the hardness and solubility of the sodium chloride crystals in ordinary table salt. When an atom of sodium joins an atom of chlorine to form salt, one of the sodium electrons jumps over to the outer shell of the chlorine atom, becoming part of it. Since the sodium has lost an electron and the chlorine has gained one, they attract each other like electrically charged conductors with a force that is difficult to overcome.

A covalent bond between two atoms of carbon accounts for the almost infinite variety of carbon compounds. A covalent single bond is formed when one electron from each of two carbon atoms combines with the other to form an electron pair which is shared by the two atoms. When two electrons from each atom are shared, a double bond is formed. Since a carbon atom has four electrons, each of which can contribute to a bond with another atom, there exist many opportunities to form rings and chains which eventually become very long and grow into giant molecules.

Weighing a molecule

While the quantum theory answered vital questions about the structure of individual atoms and of small molecules, there remained the puzzle that had incited animated arguments among chemists for almost a century: why are giant molecules such as cellulose, starch, proteins and rubber so different in nearly all characteristics from small ones, even though their chemical composition is the same or at least very similar? So long as there was no reliable method of measuring the high molecular weight of these substances, there was not much hope for a scientific approach to the problem. In the early 1920s the Swedish chemist and physicist The Svedberg provided the first dependable method of determining the weight of giant molecules with the aid of a sophisticated cream separator—the high-speed centrifuge.

For years scientists had used the centrifuge as a source of artificial gravity to separate the light and heavy particles in a suspension—just as a farmer uses a simpler centrifuge to divide light cream from heavy milk by whirling the liquid at a high speed. The centrifugal force throws the heavier particles farther from the axis of rotation than the lighter. But Svedberg wanted to learn the relative weight of certain large molecules. These differed very little in weight—they were all quite heavy—and to distinguish these slight variations he needed a centrifuge with enormous speed. By using a rotor driven by an oil turbine, he was able to increase the speed of the centrifuge to some 60,000 revolutions per minute—faster than the turbine of a modern jet airplane—thus providing centrifugal force 250,000 times the pull of gravity.

With this tremendous force he was able to measure the settling, or

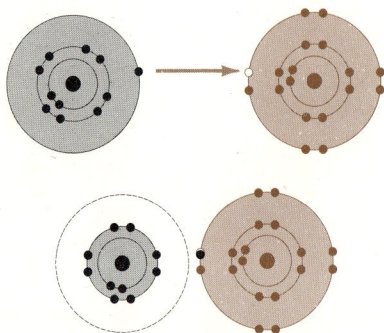

AN IONIC BOND joins atoms of sodium and chlorine *(top)* in sodium chloride, or common table salt *(bottom)*. When apart, both sodium and chlorine atoms are electrically neutral —each has a positively charged nucleus balanced by the negatively charged electrons surrounding it. When the atoms meet, chlorine "steals" one of sodium's electrons, thereby converting the atoms to oppositely charged ions. A strong electrical attraction between the ions provides the bond which holds the sodium chloride compound together.

sedimentation, rates of many molecules. From these findings he calculated the molecular weights of the molecules. Organic molecules were indeed giants among molecules, marked by weights thousands of times greater than those of other molecules. In 1926, Svedberg was awarded the Nobel Prize in chemistry for this work.

Combining this information with the new theories of atomic and molecular structure, the German organic chemist Hermann Staudinger inferred in the 1920s that a major difference between ordinary and large molecules must be simply their size. He advanced the idea of the giant molecule, whose properties differ from those of ordinary substances only by reason of molecular weight and structure. Staudinger ventured further. Since nonliving molecules are small and living ones large, he maintained that size again was the only distinction between them—and he went so far as to suggest that life itself somehow emerges whenever molecules of sufficient size and complexity are assembled.

Staudinger's conclusions had been foreshadowed by Hermann Emil Fischer who, at the turn of the century, had shown that highly complex proteins—the principal constituents of living cells—were made by the linking of the relatively simply structured amino acids; he had actually synthesized a protein fragment with a molecular weight of about 1,500. (The lightest molecule, that of hydrogen, has a molecular weight of two.)

As additional proof of his contention, Staudinger converted cellulose, the main structural component of plants and trees, into a completely different substance, cellulose acetate, and then reconstituted it by purely chemical means. He thereby showed that a giant molecule derived from a living process could be changed and then reproduced by purely chemical processes.

A quarrel between scientists

Staudinger's theory that life itself could come spontaneously from giant molecules evoked a 20th Century version of a medieval theological dispute. In technical journals and scientific meetings, Staudinger's opponents attacked his thesis as lacking foundation, and offered their own conjecture that there might be some unknown form of bonding, some undiscovered life force, that holds the giant molecule together. If Staudinger was correct, they said, it would be possible to create life in a test tube.

On the basis of his experimental evidence, Staudinger maintained that he could not find anything mysterious or unorthodox in the bonding of giant molecules. It was based, he said, on simple organic chemical principles; therefore, normal organic molecules, if sufficiently large and complex, must somehow be the source of life. In the light of all later findings, the dispute has now been settled in favor of Staudinger.

After this theoretical groundwork had been laid and with the new tools available, the science of polymer chemistry had become fully established, and because of its challenge it now attracted some of the most brilliant of the younger generation of chemists. The challenge was a double one:

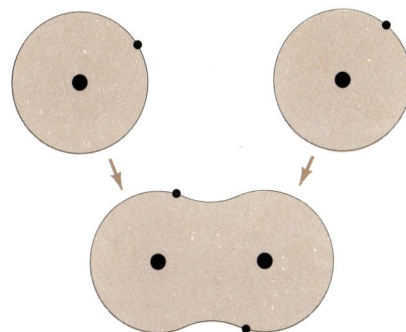

A COVALENT BOND is formed between hydrogen atoms *(top)* when they share a pair of electrons. Unlike sodium and chlorine, which are oppositely charged when they are combined *(opposite)*, hydrogen atoms remain electrically neutral even after they have joined forces to make a molecule. The atoms merely draw close enough together so that their electrons begin orbiting around both nuclei *(bottom)*. The strong bond resulting from this sharing of electrons is the type that holds most giant molecules together.

first, to establish the graphic formulas of natural polymers like protein, cellulose and rubber, and by so doing to arrive at methods for synthesizing them; second, to construct giant molecules which did not occur in nature and which might, like Bakelite, have useful properties unknown in any natural material.

Inevitably, the possibilities of polymer chemistry aroused a great deal of attention in the chemical industry. Bakelite had already proved a resounding commercial success, and it was possible that other polymers as yet unknown might have even greater potential. The large chemical firms began to look for scientists trained in the theoretical and experimental foundations of polymer science, who could pioneer in the promising field.

In 1927, Germany's giant I.G. Farben Company recruited a staff of more than 20 scientists (including Herman F. Mark, author of this book) for a polymer research laboratory. The efforts of this group were spectacularly successful in developing a sound theoretical basis for the structure of giant molecules and practical ways for their synthesis. The Farben chemists refined the original, qualitative concepts of polymer structure so that they could plan with pencil and paper the individual steps which yield new giant molecules from a wide variety of starting materials. As a result, the group began to turn out a flood of new polymers in the laboratory, many of them of great importance.

The first of the new polymers to show commercial promise was polystyrene, which Farben began producing on a mass scale in 1929. Polystyrene is built up by linking together in a long chain the small molecules of styrene, a substance easily prepared from the hydrocarbons of petroleum or coal.

Polystyrene proved to be a useful material in itself, since it is water-clear, hard and shiny, can be molded at the comparatively low temperature of about 250° F., and objects made of it readily assume a whole rainbow of translucent, transparent or opaque colors. But even more important from an industrial point of view is the fact that styrene can be combined with other molecules into copolymers, i.e., giant molecules made by the joining together of two different kinds of small molecules. As will be shown, the combination of butadiene, another hydrocarbon derived from coal or petroleum, with styrene produced an excellent synthetic rubber.

One new product per day

In the period between 1929, when the I.G. Farben research team produced the first polystyrene, and 1932, the group turned out synthetic polymers at the rate of about one a day. Not all of them were useful, of course, but some were of great commercial value. Among these were the first polyvinyls, used for floor tiles and other purposes, and the polyacrylics, some of which later were used to make excellent fabrics such as Orlon and Acrilan, and hard, clear plastics like Plexiglas. These alone could form the basis for a sizable plastics industry.

It was during the 1920s that the term "plastics" became popular to describe certain commercial products resulting from polymer research. The word, intended to indicate materials that could be shaped or formed, now covers a bewildering number of materials, ranging from blood-plasma substitutes and stocking yarn to the tough timing gears of automobile engines. The variety of uses is enormous. However, the situation is not confusing once the plastics are sorted out according to their varying structures and properties.

There are two major kinds of plastic polymers. One is called "thermoplastic," indicating that the materials soften on heating and harden on cooling in a virtually reversible manner; they are molded by extrusion, injection or compression, and are used to make fibers, raincoats, ice buckets and thousands of other products. Their giant molecules have no strong links between the individual molecules; these substances, therefore, can be heated and molded over and over again.

Building a plastic

The other big family is described as "thermosetting," and is made up of substances which polymerize irreversibly under heat or pressure, sometimes with, sometimes without catalysts, to form a hard, rigid, infusible mass. The building units of thermosetting plastics—small organic molecules—are formed as powder, pellets or viscous liquids. The material is put in a heated mold, and polymerizes to its final shape, which may be used in anything from costume jewelry to a TV cabinet, from a foam cushion to a boat hull.

Structurally, thermosetting plastics are characterized by the cross-linking of long polymer chains through strong chemical bonds which make the material insoluble, infusible, hard and rigid. However, modern techniques make it possible to vary the frequency of the cross-links between the long chains, to produce polymers that range from tough, flexible substances such as epoxy resins in glues and paints—which have few intermolecular bonds—to hard, dense solids such as Melmac, with many cross-links, for use in cups, plates and similar products.

Shortly after I.G. Farben decided to gamble on pure research in the polymer field, Du Pont, the largest chemical company in the United States, made a similar decision. The chemist they chose to head the project was Wallace Hume Carothers.

In 1928, when he went to work for Du Pont, Carothers was 32 years old. He was one of the first chemists to try to apply the concepts of quantum mechanics to the understanding of chemical bonding in organic molecules. In a paper on "The Double Bond," which he wrote as a graduate student, he developed the theory of the covalent bond virtually to the point where it stands today.

When Carothers was placed in charge of a team of highly skilled chemists, Du Pont embarked on a new type of long-range research with no expectation of immediate practical results. Under the circumstances it was hardly surprising that the team's new research laboratory, devoted entirely to pure science, was known as "Purity Hall" around the Du Pont Experiment Station.

Carothers knew exactly what he was after, and spelled it out in a letter to a chemist friend: "One of the problems I am going to start work on has to do with substances of high molecular weight. I want to attack this problem from the synthetic side. One part would be to synthesize compounds of high molecular weight and known constitution."

The first objective of the Carothers team was to study the mechanisms of various polymerization reactions. Over the next decade the group turned out more than 30 scientific papers, laying most of the groundwork in this field. One mechanism is condensation polymerization, by which two organic small molecules link and form a new larger molecule plus water. Observing a condensation reaction in 1930, Julian Hill, one of Carothers' associates, produced a thick syrup in his beaker. When he took a glass stirring rod out of the beaker, long threads of the syrup were drawn along with it. These threads, instead of breaking or dripping away, solidified on cooling, stretched like rubber, but when released did not snap back but became strong, tough, lustrous and somewhat elastic. As Carothers and Hill reported at a scientific conference, these fibers, emerging as an accidental by-product of a large fundamental research project, demonstrated "for the first time the possibility of obtaining useful fibers from strictly synthetic materials." What had happened, as Carothers immediately recognized, was that stretching had pulled the giant molecules into long parallel bundles that were equivalent to those of natural fibers like silk or cellulose.

A historic thread

This laboratory thread was a true synthetic fiber, the world's first, built up into a long continuous chemical structure from organic and inorganic raw materials. Artificial silks and rayons, which had been produced for many years, were not true synthetics. They depended for their properties on the cellulose from which they were produced.

Purity Hall's first fiber demonstrated the possibility of synthetics but was useless in itself; it was comparatively weak and had a low melting point. For the next four years the laboratory tried thousands of chemical combinations, and produced hundreds of different fibers, all deficient in one way or another.

From the point of view of Carothers' employers, these years were far from wasted, for in two side excursions from his continuing fiber research, he laid the basic groundwork for the polymerization of chloroprene to produce neoprene, one of the more useful synthetic rubbers, and developed a synthetic musk which was sold to the perfume trade.

In 1934 Carothers squeezed out of a hypodermic needle a thread of

A NYLON MOLECULE is a long chain of repeating units made from adipic acid and hexamethylenediamine. Chemists treat the adipic acid molecule *(right)* so that it loses one hydrogen atom and one oxygen atom, and the hexamethylenediamine so that it loses one hydrogen atom *(box)*; the cast-off atoms join to form water (H_2O). The two molecules also join to start a nylon chain, shown on the opposite page. Only a small part of the nylon polymer is shown here—and although the complete chain has some 1,700 atoms, it is only an average-sized polymer.

ADIPIC ACID   HEXAMETHYLENEDIAMINE

the first fully practical synthetic fiber—which Du Pont was to call "nylon." It was a product of hexamethylenediamine and adipic acid, both of which can be made from carbonic acid. This was only one of a series of similar polymer fibers developed by the Carothers team, but it was the one Du Pont decided to put into production first. No one ever had cause to question the wisdom of that decision.

In the original patents on nylon, granted in 1937, many claims were made, among them that the "new artificial silk" was stronger than natural silk or rayon, and that a strand could be drawn to a size much finer than that of a natural silk fiber. The synthetic, it was said, was also more elastic and could be drawn to as much as 700 per cent with good recovery.

An ironic triumph

With the granting of these patents, Carothers' research program might be said to have been crowned with complete success. Du Pont had rushed through the design and building of the first factory for nylon fiber. But Carothers did not live to savor his triumph. Although he was a man of strong constitution and brilliant mind, he had suffered from increasing fits of depression, and in the spring of 1937 he committed suicide.

Carothers' death was an ironic counterpoint to the glamorous career of his brainchild. By the time nylon went into production, the company had spent some $20 million in research and development. There were many observers who felt that this was a heavy gamble on an untried product. Others predicted freely that no synthetic fiber could ever be used for more than blending with natural fibers.

But, thanks to the exceptional properties of nylon and, in part, to skillful advance publicity, women stood in long lines to buy the new stockings "made of coal, air and water" when they went on sale in May 1940. Most of the initial supply of four million pairs was gone in four days. Experienced merchandisers were astounded by nylon's success.

Nylon today is still the most used of the synthetic fibers. It goes into the sails of racing yachts, into hairbrushes and parachutes, and into various kinds of bearings; in some ways it remains unmatched by the other synthetic polymer fibers which have followed it. Production has risen from 50,000 pounds in 1938 to close to 1,400 million pounds annually, in the United States alone.

The wide and almost instant public acceptance of nylon had one far-reaching effect. Previously the man in the street had been skeptical about synthetic products, feeling that they were basically cheap substitutes for the superior natural materials. After nylon he was ready to accept synthetics on their own merits as the peers and often the superiors of natural materials. In terms of acceptance, the polymers had come of age.

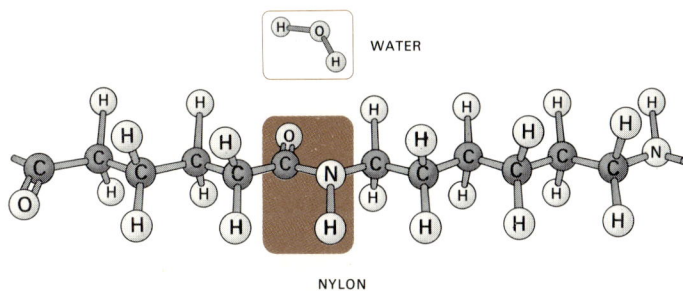

WATER

NYLON

The Colossus
of Chemistry

In the spring of 1804, when the first gunpowder from his Delaware mills went on public sale, Eleuthère Irénée du Pont thought his new business might someday make as much as $10,000 a year. The 33-year-old French political refugee proved a better powder maker than prophet. By 1811 the Du Pont Company was the largest powder-making firm in America. The development of synthetic explosives later in the century provided two more profitable products, dynamite and smokeless powder—which gave the company the chemical knowledge needed to produce its first synthetics.

This sideline became the main line after the First World War, when Du Pont used its profits to buy up a number of existing chemical companies, as well as the rights to such promising substances as cellophane—a big moneymaker since 1927. The company's own research produced both neoprene (right) and nylon in the 1930s, and helped make Du Pont the world's largest manufacturer of chemical products. Today more than 20,000 products pour out of its 147 plants each year. Irénée du Pont would be astonished at their total value: almost three billion dollars.

A CREATOR OF SYNTHETICS
Wallace Hume Carothers displays the merits of neoprene, the first commercially successful synthetic rubber, which he helped create in 1931. Carothers was a brilliant young chemist at Harvard when Du Pont hired him in 1927 and gave him a free hand to investigate any synthetic substances he chose. Before he died in 1937 he had directed the research which eventuated in nylon.

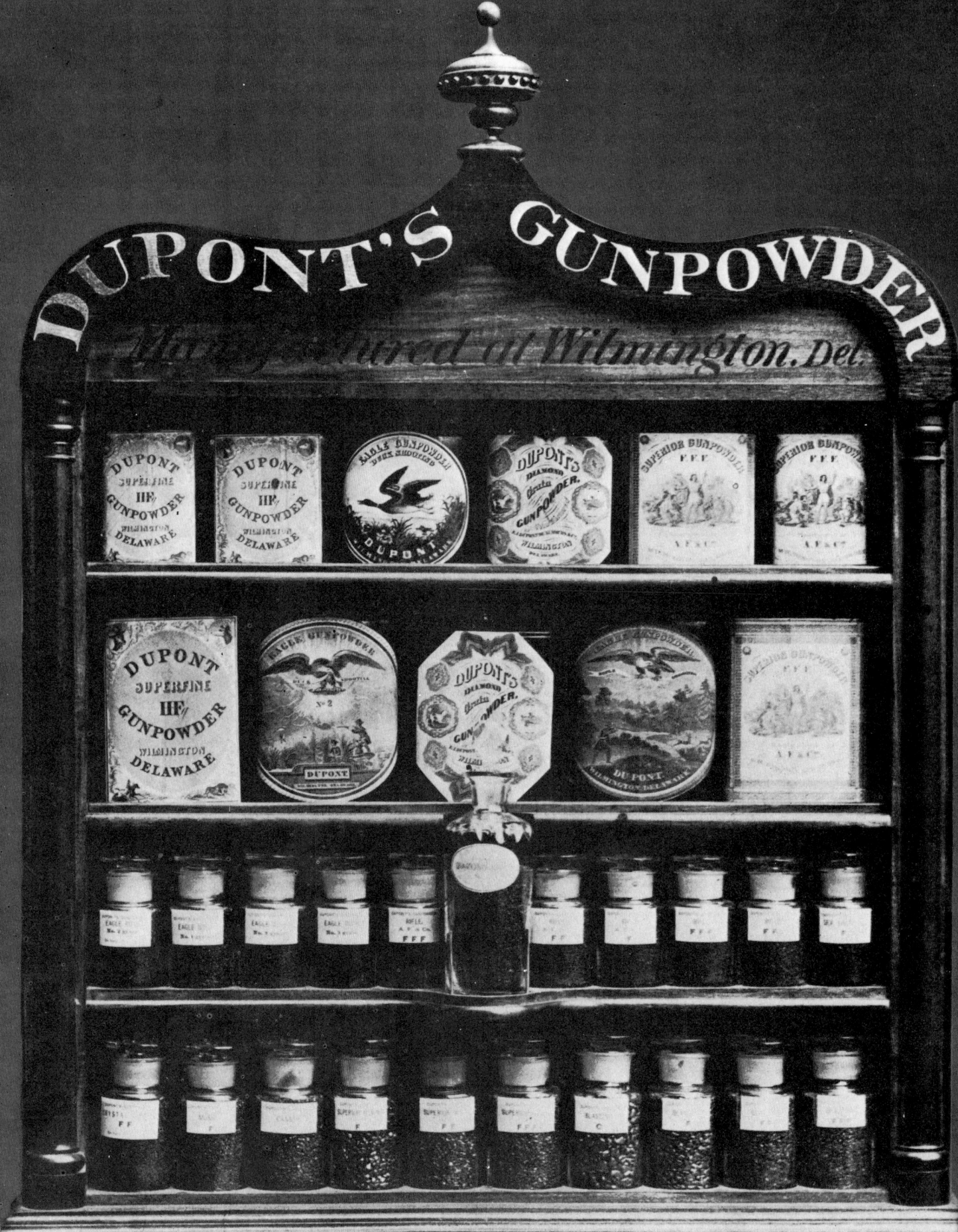

By the 1870s Du Pont's lethal showcase included 21 grades of powder. Finer grains were for guns, the coarse powder for blasting.

Gunpowder for a Young Nation

Explosive powder was an invaluable commodity in the early years of the United States. Americans used it for many purposes—to fight off Indians in upstate New York, to blast stumps off farmland in Virginia, to mine iron in Pennsylvania, to quarry granite in Vermont. And no powder was better than that made by Eleuthère Irénée du Pont in his mills by the banks of Delaware's Brandywine Creek.

Du Pont had learned the recipe for the manufacture of powder from the great French chemist Antoine Lavoisier—75 parts of saltpeter from India, added to 15 parts of sulfur from Sicily and 10 parts of charcoal, made from burnt willow twigs. In 1804, the company brought out an all-purpose black powder in several grains—some for guns, others for blasting. But as the mills expanded (bottom), so did the Du Pont line, which by the 1870s contained the assortment opposite.

The first Du Pont powder mills were built in 1802 on a wooded 95-acre site near Wilmington, Delaware. This is Du Pont in 1840.

A HAZARDOUS INDUSTRY

Gunpowder-making was a dangerous business. The early Du Pont buildings were separated to ensure against the destruction of the entire plant by a single explosion. Each structure had three heavy stone walls; the fourth wall, of wood, faced the creek. Eleuthère Irénée du Pont built them that way so that the force of an explosion would be dissipated in the least dangerous direction. Furthermore, no spark-producing metal tools were allowed inside the mills. Despite these and other precautions by Du Pont, two early gunpowder explosions cost 45 lives.

The Power of Chemistry

A 145-ton blast of dynamite clears a rocky reef from New York City's East River in 1885.

The spaghettilike strands of nitrocellulose being extruded out of Du Pont's smokeless-powder machines during World War I *(right)* signified the end of one era in the history of the company and the beginning of another. For nitrocellulose—cotton treated with nitric and sulfuric acids—was not only far superior to gunpowder, but could also be transformed into new cellulose products, such as celluloid.

Whereas making gunpowder had been a simple matter of mixing a few ingredients, smokeless powder involved a chemical process. Along with another chemically produced explosive—dynamite, made with nitroglycerin *(left)*—smokeless powder led Du Pont into the related field of plastics.

AN EXPLOSIVE EXPANSION

The bustling Du Pont mills of 1854 turned out much of the black powder used in the Crimean War. During the Civil War, Du Pont supplied half of the Union's gunpowder—four million pounds—and emerged at the end of the war as one of the world's largest powder companies. Some of the buildings shown in this old print are preserved as a historic industrial site.

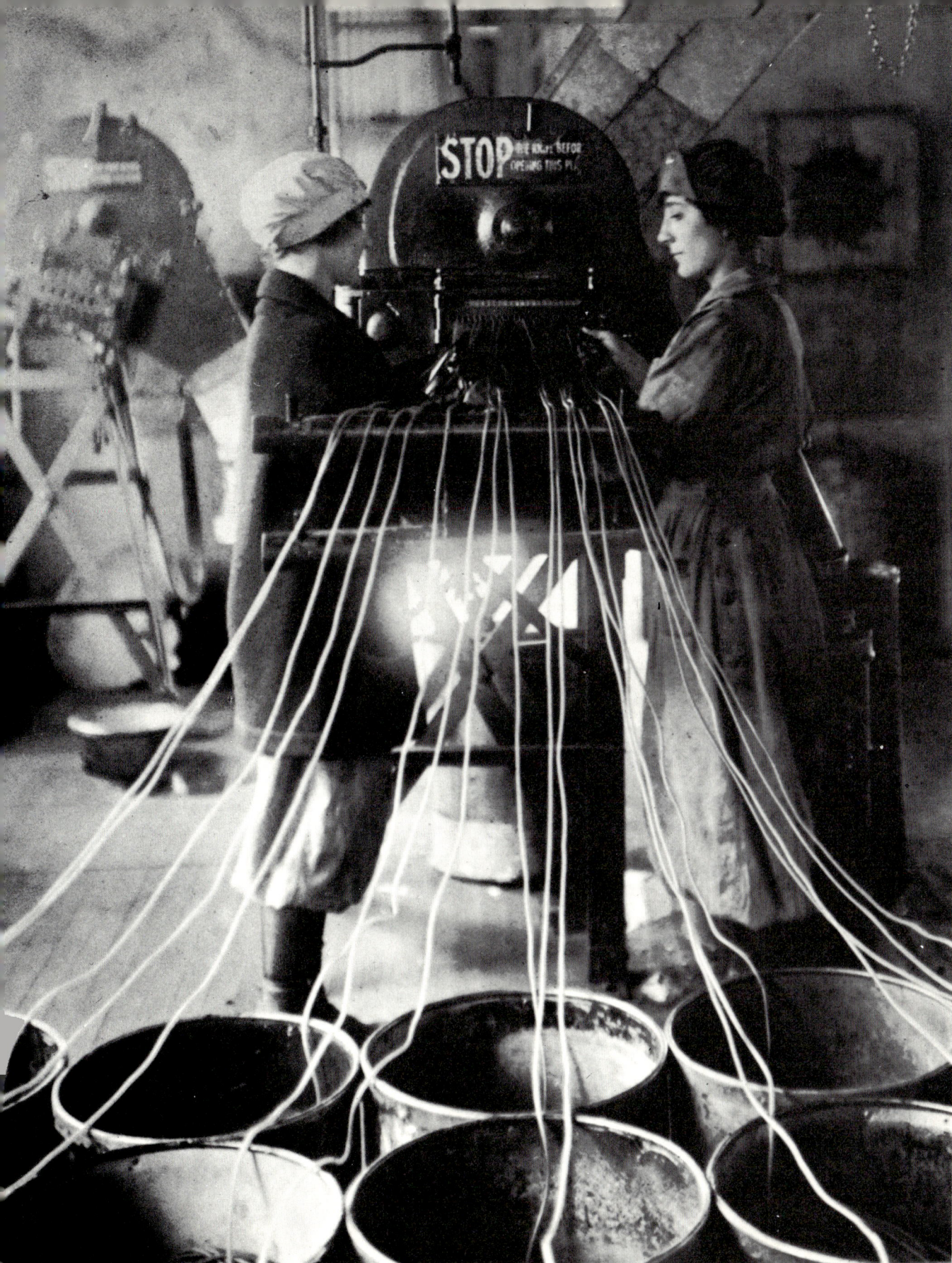

DEADLY STRINGS OF POWDER

Ribbons of explosive nitrocellulose like these—so safe to handle that women were hired to tend the machinery—provided much of the fire-power for World War I. Cut up and loaded into cartridges, nitrocellulose became smokeless powder, far more powerful and efficient than gunpowder. Du Pont supplied the Allies with 40 per cent of all explosives used in the War.

Queen Margherita's Limousine
(QUEEN DOWAGER OF ITALY)

Is Upholstered with

DU PONT FABRIKOID
REG. U S. PAT. OFF.
MOTOR QUALITY

Having acquired the Fabrikoid Company in 1910, Du Pont lost no time in advertising its artificial leather.

DU PONT AMERICAN INDUSTRIES

First Aid
for Breaks and Tears

Save your household treasures. Be they of marble, ivory, pearl, metal, china, wood, leather, py-ra-lin or glass, Du Pont Household Cement will make them whole again.

Du Pont Household Cement is a pyroxylin cement made of pure cotton (no animal refuse). Waterproof, transparent, does not become rancid or harden in the tube. A necessity in every home.

DU PONT CHEMICAL WORKS
WILMINGTON, DELAWARE

Send for illustrated folder giving its many uses and name of the nearest dealer, or a full sized sample tube sent, postpaid, to any address on receipt of 30c in stamps.

DU PONT HOUSEHOLD CEMENT — WATERPROOF TRANSPARENT — MENDS PERFECTLY

DU PONT

DU PONT AMERICAN INDUSTRIES

Loveliest for the Bride

IVORY PY-RA-LIN

Even the rarest orchid or fairest porcelain cannot surpass Ivory Py-ra-lin in beauty. The bride, of all persons, appreciates the simple elegance of such a gift.

Ivory Py-ra-lin is the present-day manufacturing expression of an "ivory" caste in toiletware that has come down through the ages. Its exquisiteness and usefulness endure.

Ask one of the better dealers in your town to show you his Ivory Py-ra-lin exhibit. Particularly the pattern Du Barry. Every piece of the genuine merchandise is plainly stamped with the trade-name Ivory Py-ra-lin.

THE ARLINGTON WORKS
OWNED AND OPERATED BY
E. I. DU PONT DE NEMOURS & CO.
725 Broadway, New York

Canadian Office and Factory, Toronto, Ont.   Boston, St. Louis, Chicago, San Francisco

DU PONT AMERICAN INDUSTRIES

E. I. Du Pont De Nemours & Co.   Wilmington, Del.   The Arlington Works.   725 Broadway, New York
  Industrial, Agricultural, Sporting and Military Explosives   Owned and Operated by E. I. Du Pont De Nemours & Co.
Du Pont Fabrikoid Co.   Wilmington, Del.   Ivory Py-ra-lin, Auto Menning, Challenge Cleanable Collars
  Leather Substitutes
Du Pont Chemical Works.   170 Broadway, New York   Harrison's Inc.   Philadelphia, Pa.
  Pyroxylin and Coal Tar Derivatives   Paints, Pigments and Chemicals

DU PONT

TWO USES OF PYROXYLIN: A LIQUID CEMENT STILL SOLD TODAY . . .   . . . AND AN IVORYLIKE PLASTIC. THESE ADS DATE FROM AROUND WORLD WAR I.

The Turn toward Plastics

At the beginning of its second century, Du Pont took a calculated gamble: it surrendered its world leadership in explosives and placed its faith in the newfangled field of plastics. The oval Du Pont trademark soon appeared on celluloid collars that could be wiped clean instead of laundered, and on a host of products based on pyroxylin—a nonexplosive version of the same nitrocellulose formula used in smokeless powders. This was a versatile synthetic: it made household cement that could mend fine china, converted cotton cloth into a fabric that looked and felt like leather, and could be molded into a plastic resembling ivory.

After 1918, Du Pont expanded still further into plastics by purchasing smaller companies and adding their products to the Du Pont line. It also used its war-learned knowledge of coal-tar chemistry, previously dominated by Germany, to establish an entirely new line of synthetic dyes, acids and paints.

As early as 1902, Du Pont had set up a research laboratory to find new applications for chemicals; by 1906 the research budget was $300,000 a year—a staggering sum at the time.

Throughout the 1920s the laboratory produced a steady flow of synthetic resins and lacquers, and was responsible for the success of cellophane. This transparent cellulose film was invented by a Swiss chemist, Jacques Brandenberger, in 1912, and had enjoyed mild success in Europe, mostly as a wrapping for perfume bottles. Sensing cellophane's potential, Du Pont bought the U.S. patent rights in 1923 and by 1927 had devised a way to moisture-proof it. Cellophane replaced wax paper as the wrapping for perishables, and almost overnight it became an American institution.

1915: CELLULOID COLLARS FOR "EVERY WALK OF LIFE."

1927: A WONDERFUL WINDOW OF MOISTURE-PROOF CELLOPHANE.

Nylon: The Miracle Fiber

The giant molecule polyhexamethyl-eneadipamide—better known as nylon—was not born easily. As early as 1930, a Du Pont chemist, Wallace Hume Carothers, was convinced that a durable fiber could be created from molecules derived from coal, air and water. But it took 10 years and $27 million to perfect the process.

In 1940, its first year on the market, nylon was a sensation: it went into toothbrush bristles, surgical sutures, fishing line, but most of all into women's stockings—64 million pairs. The next year nylon went to war, and Du Pont turned out enough yarn to make four million parachutes and enough tent fabric to cover Manhattan. Today, nylon is still Du Pont's most popular product, accounting for $500 million in sales every year.

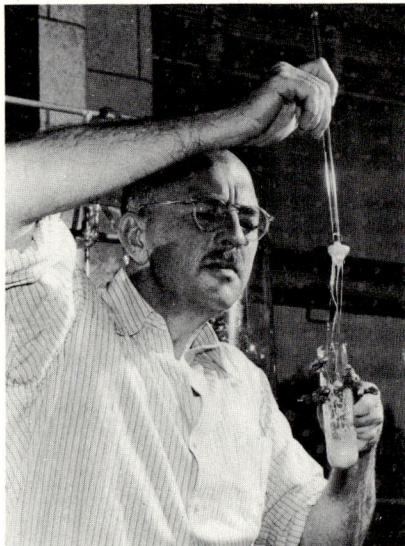

THE BIRTH OF NYLON
Julian Hill, Carothers' chief assistant, re-creates the moment in 1930 when the first truly synthetic fiber was found. A durable version of the fiber—nylon 66—was not created until 1935.

NYLON BY THE MILE
Custom-designed machinery—guarded by more than 200 Du Pont patents—had to be built to mass-produce nylon. This early unit, now an antique, pressed nylon filaments into a long ribbon, which was later cut, melted and extruded to produce nylon fibers of various thicknesses.

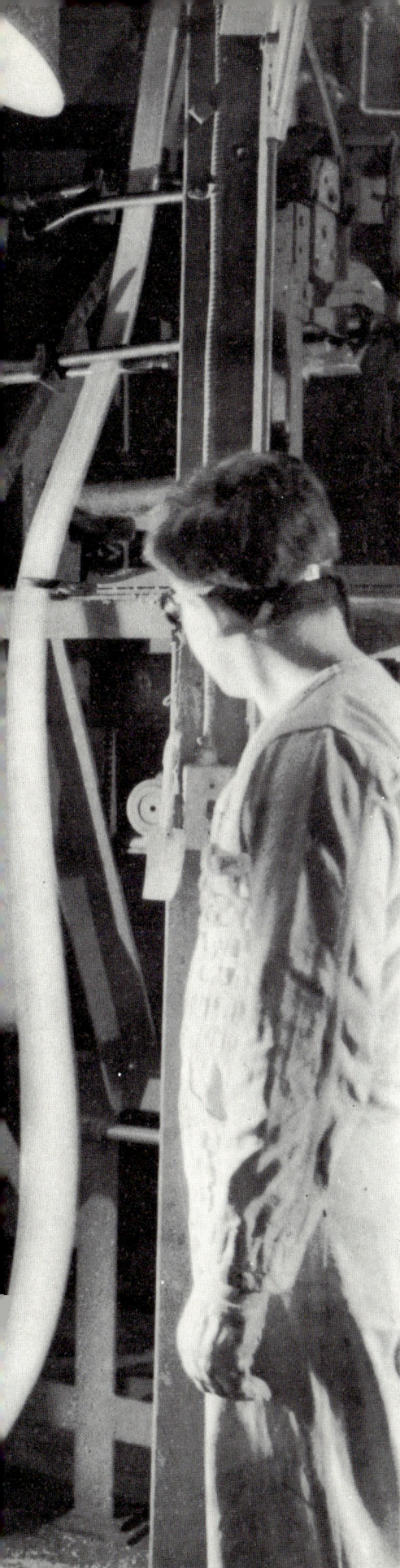

NYLON GOES TO WAR . . .

Nylon parachutes, stronger and longer lasting than those of silk, played an important role in World War II. Cut off from supplies of Japanese silk, the U.S. armed services quickly adopted Du Pont's new synthetic as a superior replacement. A total of 51 million pounds of nylon went into military parachutes and cord for airplane tires.

. . . AND COMES HOME AGAIN

A young woman, oblivious of onlookers in a postwar "nylon line," tries on her stockings outside the store. Although 90 per cent of all nylon went to hosiery manufacturers after V-J Day, it took two years—and 700 million pairs of stockings—before the supply equaled the demand.

A SIEVE OF HOLLOW FIBERS
Fine, hollow, nylonlike fibers, shown at right magnified 250 times, are used to filter water in a new kind of compact purification unit. Above, a technician looks on as the fibers are laid down on porous material, which will then be rolled up and encased in a cylinder. In use, brackish water is pumped in at about 400 pounds per square inch and flows around the cylinder's approximately one million tiny fibers. The fiber walls block the minerals but permit the pure water to dribble into the holes to be collected.

A Booming Brood
of Synthetics

Hundreds of useful synthetics have poured out of Du Pont's laboratories since nylon. Some of the more recent ones include Mylar, a transparent plastic film as thin as a human hair but strong enough to lift an automobile; Surlyn, an improved and transparent version of polyethylene that can be used to make see-through containers for liquids; and very special, tiny, hollow fibers for use in water purification plants.

Not all of the Du Pont synthetics have filled the role for which they were originally intended. A notable example is Orlon (below), an acrylic fiber created for outdoor uses. When the Du Pont chemists discovered that its silky strands tended to disintegrate with prolonged exposure, they adapted the fiber for use in clothing and blankets—and turned failure into success.

The flow of synthetics shows no signs of abating. Du Pont is hard at work on paints that will never chip or peel, preservatives that will keep food fresh for years without freezing, and automobile oil that can be sealed into the engine and never replaced.

A GOSSAMER VEIL OF ORLON
Parting a skein of Orlon fibers, a Du Pont technician shows its characteristic wool-like bulk. The acrylic filaments have a permanent crimp, which keeps the fibers separated. This separation produces not only the airy lightness desired in blankets and sweaters, but also gives Orlon slacks and dresses a high resistance to wrinkling. The acrylic polymer has a high strength.

119

An Endless Research Cycle

About $250 million is plowed back into research and development at Du Pont each year, most of it aimed at discovering new synthetics. Teams of researchers drawn from a pool of Du Pont's more than 4,000 scientists and technicians often labor for years to perfect a single product.

Although the Du Pont company is a comparative latecomer in the pharmaceutical industry, it has tackled one of the trickiest problems in the business—the development of antivirus drugs. One project, pursuing a cure for the common cold, is complicated by the fact that there are nearly 100 variations of the cold virus, and a useful drug would have to be broadly effective against most of them.

While Du Pont's scientists are generally optimistic about the outcome of their work, the road from chemist to consumer is, no doubt, heavily strewn with pitfalls. By Du Pont's own estimate, only one research project in 20 ever pays off with a product that is commercially successful.

BLUE-SKY RESEARCH
A Du Pont chemist reaches into a nitrogen-filled polyethylene bag to take a sample of enzymes derived from bacteria. He is investigating the remarkable ability of the enzymes to absorb nitrogen and convert it to ammonia—a base of many synthetic molecules. From this sort of fundamental research may one day come a new synthetic product or manufacturing technique.

FIGHTING THE COMMON COLD
Searching for cold remedies, a laboratory technician introduces a dye into glass dishes that contain cold viruses, cell cultures and the experimental drug. His goal is to find a substance that will keep the viruses from multiplying, and he gauges the effectiveness of each sample's blocking action by the number of cells that survive the viral onslaught to be stained by the dye.

6

Rubber:
Better than
Nature's

A natural polymer, rubber was featured around 1900 in this ad for weatherproof buggy tops.

ON DECEMBER 7, 1941, the United States was catapulted into World War II when the sleepy peace of a Hawaiian Sunday morning was shattered by the blast of Japanese bombs and torpedoes. Within minutes, most of the U.S. Pacific fleet was sunk or disabled. Less than three months later, the supposedly impregnable British naval base of Singapore was captured by Japanese forces, and soon the Dutch East Indies were also in Japanese hands.

Of the two disasters—the crippling of the Pacific fleet and the Japanese conquest of Southeast Asia—the latter was the far more serious blow to the long-range military potential of the Allies. The United States had the materials and the technological knowledge to build new ships within a comparatively short time. But about 98 per cent of the crude rubber used by the United States had come from Malaya and the Dutch East Indies, and there was no possibility of retaking them in the foreseeable future.

When this catastrophe occurred, there were about a half-million tons in U.S. stockpiles—less than a year's peacetime consumption. Once this supply was gone, how would the nation provide this vital commodity for a great military force and the supporting civilian economy? Without rubber no truck, tank or plane could move; farm and mining machinery would be useless; internal transportation would come to a halt; electrical generators could not be built. Defeat was literally inevitable unless synthetic rubber could be manufactured in great volume.

In previous years, the United States, supremely confident that Far Eastern rubber sources could never be captured by an enemy, had given scant attention to the synthetic product. Research had been minimal. Production experience had been limited to a few pilot plants which had turned out less than 7,500 tons in 1941. Now there was a need for at least 100 times that amount—and the months that history had allotted to reach this huge output were frighteningly few.

The Axis nations had no such problem. Japan, of course, had a nearly inexhaustible supply of natural rubber from the territories it had conquered. Germany, thanks to a lesson taught that nation by the Allies in World War I, was in a similarly strong position. When the Nazis had begun rearming the country in the early 1930s, they had remembered the Allied blockade of 1914-1918, which cut off Germany's imports of crude rubber and thereby hastened its capitulation. Adolf Hitler and his cohorts had been determined that history would not repeat itself. The search for good synthetic rubbers and the means of producing them in quantity had begun even before the Nazis came to power; the work was now accelerated. The German program was a success. By the time World War II began in 1939, Germany's production had already topped 20,000 tons yearly. By 1942 it was more than 90,000 tons.

To a major degree, then, the outcome of the war was dependent upon a scientific and technological race against time. It was a race involving U.S. chemists and their ability to imitate the German products, and engineers and production experts who must translate the achievements of the laboratory into the massive operations of great factories.

It was ironic that the United States should ever have faced a shortage of natural rubber. The first rubber ever seen in Europe was probably brought from the New World by Christopher Columbus in 1496, and the Western Hemisphere was the sole supplier for centuries. Among the exotics in Columbus' cargo were some oddly bouncing balls which the natives of Haiti used as toys. Later, Cortez found the same substance in Mexico. However, Europeans showed little interest in the unusual material until 1731, when Charles de la Condamine went to Peru on an earth-measuring assignment for the French Academy of Sciences and returned with large rolls of the dried extract of *Hevea brasiliensis*, the "weeping tree" of the Amazon Valley.

Condamine found that the Amazon people used caoutchouc, as it was called, for a good deal more than amusement. By spreading a thin film of the sap of the hevea on cloth, and then hardening it by exposure to sunlight or smoke, they fabricated crude waterproof garments. They also made bottles from the sap (now called latex), and even anticipated present-day expensive "molded" shoes by pouring the latex on their feet and legs and letting it dry into waterproof boots. However, Joseph Priestley, the great 18th Century English chemist, apparently ignored these significant qualities of the material—its elasticity, flexibility and impermeability—when he gave it its English name. It was "excellently adapted," he noted, "to the purpose of wiping from paper the marks of a black lead pencil," i.e., by rubbing. And because it came from the West Indies, it received the name "India rubber."

In 1823, the first important application of rubber was developed when a Scotsman named Charles Macintosh dissolved rubber in naphtha and sandwiched it between two layers of cloth, creating the raincoat fabric that still bears his name. The lamination mitigated the effects of extreme temperatures on the rubber, a fact overlooked by the first American manufacturers, who produced unlaminated imitations; the hot American summers turned their rubberized sheets into gluey capes.

An industry from an accident

Rubber remained a substance of limited utility until 1839, when Charles Goodyear discovered the process of vulcanization. Legend has it that Goodyear, a New England inventor, capped 10 years of unsuccessful experimentation with a happy accident. While searching for a way of preserving rubber, he spilled a mixture of latex and sulfur on a hot stove. He scraped it off, let it cool—and then found that it was no longer sticky, that it would spring back to its original shape after stretching or twisting, that it stayed flexible in cold temperatures and solid in hot, and that it resisted chemicals that dissolved raw rubber.

Although Goodyear did not profit personally from his discovery—after years of court battles over patents, he died in 1860 leaving his wife and six children with debts of $200,000—vulcanization made possible the enormous growth of the rubber industry. The greatest boom came about

TAPPING A RUBBER TREE, a worker *(above)* makes a V-shaped cut in the trunk to sever a network of tiny latex vessels in the bark. As shown in the magnified cross section below, these vessels crisscross through the soft inner layer of bark. Latex, a milky white fluid, drips from the cut, and is purified to make rubber. Most of the world's rubber plantations are now in Malaysia, where an acre of good trees will yield up to a ton of rubber a year.

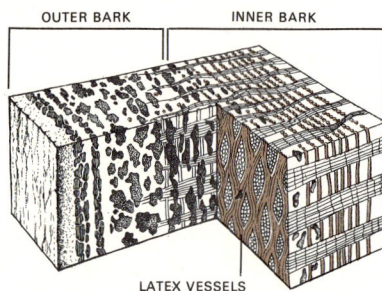

OUTER BARK INNER BARK

LATEX VESSELS

the turn of the century. The pneumatic tire, invented by Robert Thomson in 1845 but ignored for years, was reinvented by a Belfast veterinarian for his son's tricycle. After improvements by the Michelin brothers André and Edouard in France, the tire became an immediate success in the automobile age which was just getting underway.

Even before the coming of the automobile, the demand for rubber had exceeded the production of the primitive tree-tappers of the Amazon jungle. Great Britain, among other users, wanted to start hevea groves in the Far East, where conditions were much better for large-scale, businesslike operation, but the South American countries firmly opposed exporting any seeds. They had good reason: in the 1850s the Dutch East India Company had taken cinchona trees, the source of quinine, from Peru and transplanted them to Java—and within a short time, Java had a monopoly on quinine.

Britain's respectable smuggler

But the rubber market was not to be denied. In 1875, Henry Wickham, acting on behalf of the Royal Botanical Gardens at Kew near London, smuggled some 70,000 hevea seeds from the upper Amazon. Enough of them survived the voyage to England, germination at Kew and transplantation in Ceylon to start the first great rubber plantation in the Far East. In time, Southeast Asia became the supplier of 90 per cent of the world's natural rubber. When Japan conquered that part of the world in 1942, Allied leaders must have wondered if crime—especially smuggling—really does pay.

During the entire period in which rubber evolved from a natural freak into an important and versatile industrial material, chemists were busily attempting to answer the scientific questions raised by Goodyear and other experimenters. The first objective, clearly, was to determine the very nature of rubber. What did its molecule, or molecules, consist of? What was its structure? What was the relationship of this structure and substance to its physical characteristics?

Although Faraday, Dalton and others had given these questions their sporadic attention, the first important insight into the chemistry of rubber was achieved by Greville Williams in 1860, when he isolated, by distillation from latex, a clear liquid which he called isoprene. Its molecule consisted of five carbon atoms and eight hydrogen atoms. When left in air it formed a viscid liquid, then a spongy elastic mass, which, when ignited, smelled like burning rubber. The close relationship between the small molecules of isoprene and the large ones of rubber became even more obvious in 1879 when Gustave Bouchardat polymerized isoprene with hydrochloric acid and produced a rubberlike material.

In 1882, the English chemist Sir William Tilden confirmed Bouchardat's findings. He then went on to work out the chemical structure of isoprene (page 126), which demonstrated that rubber is nothing but a huge version of isoprene—a giant molecule made up of thousands of isoprene units linked together.

Two years later, when he was teaching chemistry at Mason College in Birmingham, England, Tilden made isoprene by passing turpentine vapor through a heated tube. Part of the material he used for experiments; he bottled the remainder and put it on a shelf. When, after a number of years, he looked at the stuff he had stored, he found that it had spontaneously polymerized into a crude kind of rubber. "I have tried everything I can think of as likely to promote the change but without success," he wrote. "The [natural] polymerization proceeds very slowly, occupying according to my experience several years and all attempts to hurry it result in the production, not of rubber, but of colophene, a thick, sticky oil quite useless for the purpose to which rubber is applied." By 1907, he had decided that the synthetic product could never be a commercial possibility, and he advised the industry that its future depended upon further cultivation of the rubber tree.

The race for synthetic rubber

But this gloomy view did not lend itself to the temper of the early 20th Century. Enticed by growing markets, goaded by the military, whose equipment relied increasingly on rubber-shod wheels, and dogged by the violent price fluctuations of natural rubber, Russia, England and Germany spurred their chemists to promote the synthesis of rubber, or something sufficiently like it, to commercial feasibility.

In 1910, S. V. Lebedev, a Russian chemist, turned to a simpler starting material, one much less closely related to natural rubber than isoprene. This was butadiene, a petroleum derivative that is a colorless gas under ordinary conditions. The butadiene molecule resembles isoprene in its structure but is smaller, lacking one subgroup of carbon and hydrogen atoms that isoprene has. Lebedev succeeded in polymerizing butadiene into a giant molecule. As a synthetic rubber it was a failure, but it was an important failure, for Lebedev was on the right track in suggesting the use of simple substances that could be used to make not a rubber imitation but a rubber replacement. And his choice of butadiene as a starting material was prescient. From this inexpensive gas was to arise not one, but several useful rubbers.

Success did not come within reach until the 1920s, when the work of Staudinger, Kurt Meyer, Carothers, Herman Mark and others with giant molecules persuaded chemists that they should concentrate not only on natural rubber's chemical composition but also on its molecular structure.

Rubber, it soon became clear, is as much a state of matter as a material. Its giant molecules are characterized not only by their size (the rubber polymer consists of from 1,000 to 5,000 isoprene units), but also by their irregular structure. The molecular chains are randomly coiled, like

A POLYMER OF NATURAL RUBBER, present in the latex of rubber trees, is made up of a long chain of isoprene units whose bonding pattern has been changed by catalysts in the tree. An isoprene molecule *(left)* consists of five carbon atoms. Four of these *(brown)*, which are linked by one single and two double bonds, are affected by the catalyst. In the transformation stage *(center)*, the two double bonds open and the isoprenes join end to end until thousands are assembled to create a single molecule of rubber *(right)*.

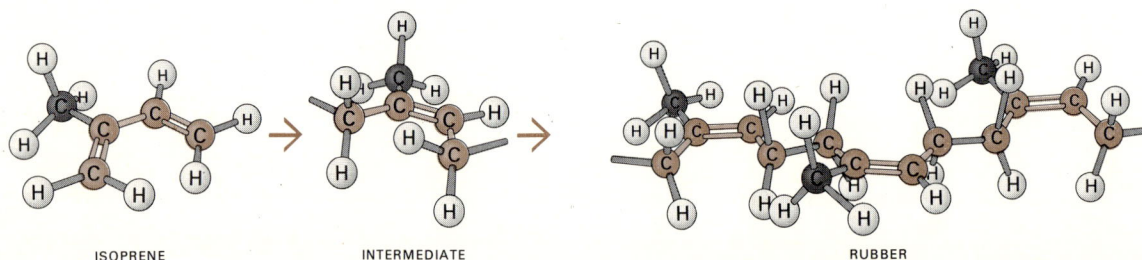

ISOPRENE   INTERMEDIATE   RUBBER

126

so much tangled cooked spaghetti. It is this irregular, or amorphous, structure that gives rubber and similar substances their resilience. When, for example, a piece of rubber is stretched, the coiled, spaghettilike molecular strands untangle; as soon as the rubber is released, the strands coil back on themselves and resume their former state.

Stretching rubber changes its molecular structure so much that its physical characteristics are transformed. The tangled strands straighten out into uniform parallel bundles. The molecule no longer has the irregular structure of soft and resilient substances; now it falls into the orderly atomic arrangement of crystalline solids such as metals. And it gains the toughness and rigidity that the crystal framework gives to solids.

Rubber's dual nature—crystalline when stretched, amorphous when not—can be seen with the naked eye. A piece of raw rubber in repose is translucent; the giant molecules are randomly coiled and represent a homogeneous mass through which light can pass. But when rubber is stretched it becomes cloudy, because the chains of molecules are pulled into tightly aligned, parallel positions like the slats of a nearly closed venetian blind, and light is scattered by them.

This molecular transposition is confirmed when unstretched rubber is frozen and then shattered: the pieces are irregular. However, if stretched rubber is frozen and shattered, the parallel strands of long fibers are clearly distinguishable.

The ability of rubber to make swift transitions back and forth between the amorphous and crystalline states is what makes it such a practical and valuable substance. In a tire, for example, the transition occurs with every turn of the wheel. The section of the tire that is bearing the weight of the car at any given moment is stretched, and therefore strengthened because its molecular structure approaches that of a solid. As the wheel turns and another section of tire assumes the weight, the tension on the first part is removed and it resumes its amorphous and resilient state, superbly fitted to absorbing road shocks.

A FANATIC INVENTOR, Charles Goodyear was driven by the belief that God had created him to cure rubber. After discovering the process of vulcanization in 1839 when he accidentally dropped a mixture of India rubber and sulfur onto a stove, he devoted his life to getting enough money to perfect the new material, finally pawning all his possessions—even his children's schoolbooks.

What makes rubber rubbery

Molecular structure alone, however, does not make rubber rubbery. Temperature is involved too. Many amorphous substances—Lucite and Plexiglas acrylates for example—become rubbery when heated. And conversely, rubber will lose its resilience and become as hard and glassy as the acrylates when it is chilled. The explanation lies in the intermolecular bonds that connect the long chain molecules. At low temperatures, the molecules contain little heat energy and move about fairly slowly. This slow movement permits many bonds to hold onto the molecules; the structure is thus kept rigid and glasslike. As rising temperature speeds up molecular agitation, the increased vibration within the molecules causes some of the bonds between them to break, and the released molecules can then move more freely. The material becomes flexible and resilient. Greater heat breaks more intermolecular bonds until the rubber is soft and gummy. When the heat is still further increased, the motion of the molecules becomes more and more violent until all bonds are broken. Then all the molecules can move independently past one another and the mass is liquefied. (This is the reason that rubber—like all amorphous substances and unlike crystalline ones —has no fixed melting point, and does not change abruptly from solid to liquid, the way ice melts suddenly into water; instead, it simply softens in a gradual transition from rigid to viscous to fluid.)

Vulcanization with sulfur inhibits, but does not eliminate the tendency of rubber to soften at high temperatures. The heat of vulcanization forces open the double bonds that interconnect atoms of the molecules. These bonds then tie onto the sulfur atoms, forming sulfur cross-links between the molecules. The additional cross-links, binding molecules with sulfur atoms, greatly strengthen the structure. With the addition of more sulfur, so many of the isoprene molecules may be cross-linked that the rubber can be converted to the rigid macromolecule of ebonite, the hard rubber of battery cases and pipe bits.

Progress toward test-tube rubber

By the mid-1920s, chemists had a good understanding of all these facts about a rubber molecule, but had yet to find the means of making one. In 1925, Father Julius A. Nieuwland of Notre Dame succeeded in making a small polymer from three molecules of the common cutting-torch fuel, acetylene, whose triple bonds made it extremely reactive, thus affording a promising start for further synthesis. Using the polymer as a base, Du Pont chemists produced a two-molecule acetylene, which, when treated with acid, yielded an even more promising material. They called it chloroprene, because its structure was like that of isoprene, except that a chlorine atom replaced one group of carbon and hydrogen atoms.

And now the well-calculated gamble Du Pont had taken by subsidizing Wallace Hume Carothers' basic research in polymerization paid an unexpected and early dividend. Carothers' group was still nearly a decade away from discovering nylon, but it had learned enough to polymerize chloroprene. The result was neoprene, the first successful synthetic rubber in the U.S., introduced in 1931.

Since its chlorine atoms gave it greater resistance than natural rubber to oils, acids, water, sunlight and oxidation, neoprene achieved immediate preference over natural rubber in hundreds of applications, including hoses, life rafts and linings for warplane fuel tanks, where it

VULCANIZING RUBBER, French workers use a huge vat in one of many plants which were built in Europe during the late 19th Century to capitalize on Charles Goodyear's 1839 invention of the vulcanization process. In such a factory, rubber was first mixed with sulfur, placed in a vat and heated with steam to convert it into a strong, permanently elastic material.

served, as natural latex does, to seal punctures. But the new rubber was too expensive for tires. And so the search for other synthetic rubbers went on.

In the early 1930s, chemists at I. G. Farben stumbled upon a laboratory curiosity. They would amuse visitors by converting isobutylene, a petrochemical, to a liquid at dry-ice temperature, and then adding a few drops of boron trifluoride, a powerful catalyst for polymerization. As soon as the drops hit the surface of the liquid, there would be a noiseless explosion and a white, rubbery snowball would form and bulge spectacularly over the edge of the glass. However, the Germans could find no way of vulcanizing the substance. Some years later, American chemists learned that the addition of butadiene—a plentiful, inexpensive gas —would remedy this shortcoming. The resulting butyl rubber is strong, elastic and so impermeable to air that it serves especially well for inner tubes, tubeless tires and similar products.

Although German chemists in the 1930s were not aware of butadiene's capacity to make butyl rubber, they had been experimenting with the gas since 1925. Their goal was to create an all-purpose rubber for tires. Their first Buna rubber (*Bu* for butadiene, *na* for the chemical symbol of the sodium used in polymerization) was appreciably poorer than natural rubber. But after further experimentation, in 1928 they hit upon a copolymer, a giant molecule made by joining two different molecules in a repeating pattern. To butadiene the Germans added still another of the many simple substances derived from petroleum—styrene, which had been the starting point for the Farben group's first great plastic success, polystyrene, a decade earlier. Combined into the copolymer Buna S (*below*), butadiene and styrene formed the first all-purpose synthetic that could economically replace natural rubber in automobile tires.

Science nullifies a blockade

This new material was so successful that it made the Allied blockade on natural rubber a matter of no military importance to the Nazis. They also conserved their limited supply of petroleum by using coal and limestone from the Saar Basin as a source of butadiene. By 1943, German production of this man-made rubber reached well over 100,000 tons annually—the equivalent of a year's yield from 40 million trees on 400,000 acres of rubber plantations.

In contrast, the United States had only the meager output from a few pilot plants when the Japanese cut off the supply of natural rubber in early 1942. Furthermore, the kinds of rubber being made had only limited uses, and the production of tires was not one of them. In August of that year, President Franklin Roosevelt dropped the rubber problem into the collective lap of a committee headed by Bernard M. Baruch.

A SYNTHETIC RUBBER MOLECULE, of a type called Buna S *(below, right),* contains two smaller petroleum products, butadiene *(far left)* and styrene *(left, brown).* In the process of polymerization, the double bonds of the styrene and butadiene molecules are broken, causing the carbon atoms to link up. In the small segment of the Buna S molecule shown below, two styrene molecules have been linked by a butadiene molecule, and each has another butadiene attached. Like natural rubber, the Buna S giant molecule, used in tires and shoe soles, actually contains thousands of its two constituent molecules.

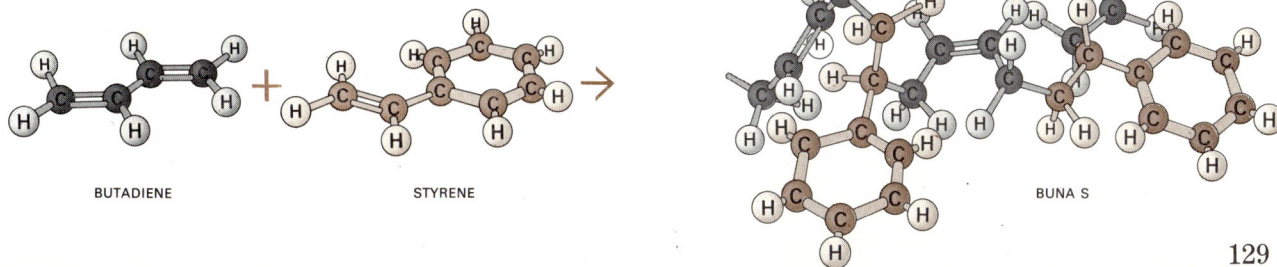

BUTADIENE   STYRENE   BUNA S

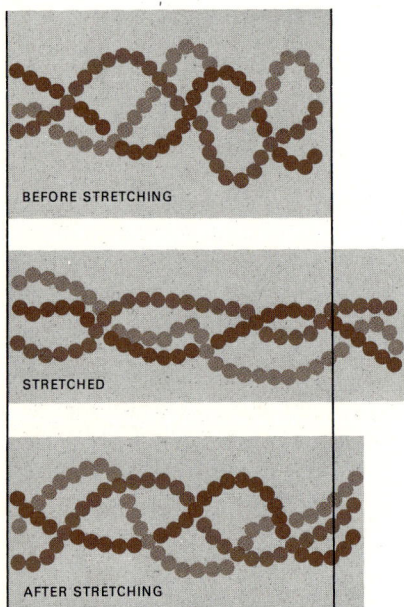

STRETCHED UNVULCANIZED RUBBER
does not return to its original state, as these
diagrams show. The original substance *(top)*
is shown here as three tangled molecular
chains, which straighten out as the rubber is
stretched *(center)*. Because the individual
chains are not connected, as they are in
vulcanized rubber, the chains slide past each
other and cannot spring back all the way to
their original positions when released *(bottom)*.

The committee's first report, issued a month later, accurately summed up the desperate situation:

"Of all the critical and strategic materials, rubber is the one which presents the greatest threat to the safety of our nation and the success of the Allied cause. Production of steel, copper, aluminum, alloys or aviation gasoline may be inadequate to prosecute the war as rapidly and effectively as we could wish, but at the worst we still are assured of sufficient supplies of these items to operate our armed forces on a very powerful scale. But if we fail to secure quickly a large new rubber supply, our war effort and our domestic economy both will collapse."

On recommendation of the Baruch committee, the President immediately issued an executive order creating the Office of Rubber Director. William M. Jeffers, President of the Union Pacific Railroad, was given the job, with a blank check to carry out the assignment, and authority which cut across the jurisdiction of all other government agencies. The race against time was on.

The first vital step in the synthetic rubber program had been taken some months earlier in a court of law, when the U.S. Justice Department managed to free the Buna S patents from the German-headed cartel that had controlled them. But a huge and complex new industry had to be started almost from scratch to supply the butadiene and styrene necessary for the manufacture of GR-S (Government Rubber Styrene), the U.S. wartime name for Buna S.

Girding for technological war

With manpower and materials of all kinds in short supply, the Rubber Director needed all the authority he had been delegated to create his new industry. Understandably, military men and civilian officials concerned with other aspects of the war effort were in constant, intense competition with the rubber program for scarce commodities.

Meanwhile, the rubber companies were pooling heretofore jealously guarded manufacturing secrets. Although the experience of any one firm in making synthetic rubber was limited, the collective knowledge of all was considerable. Contracts were made with universities and private research institutes to work cooperatively with industry. Farmers were involved too. Because butadiene and styrene can be made from alcohol as well as from petroleum, the United States used its enormous output of grain, potatoes, molasses and other carbohydrates to manufacture alcohol for rubber production and thus conserve the oil supply.

More than 50 plants were built during the next several years in California and Connecticut, in Kentucky and Texas, in Ohio, Pennsylvania and other parts of the country. Nearly half of these installations were devoted to the production of butadiene and styrene, the major ingredients of synthetic rubber; another one third of the plants manufactured the finished product. Four more factories were set up to make neoprene, butyl and Thiokol, varieties of rubber needed in lesser quantities. Nine plants were erected to make alcohol and other needed chemicals.

The results of all this collaborative effort soon became manifest. In 1942, the first desperate year, production of synthetic rubber was slightly more than 20,000 tons. The following year, output had increased more than tenfold to nearly 215,000 tons, about twice what Germany made in the same year. In 1944, when the Allies' great counteroffensives were launched in both the European and Pacific theaters, the U.S. turned out more than 700,000 tons, and in 1945, 755,000 tons.

A billion-dollar solution

The solution to the potentially disastrous rubber problem had cost about a billion dollars. But the solution was found and it was one of the great technological triumphs of history. It not only made the Allied victory possible, it also thrust the U.S. into the leadership of the synthetic rubber industry throughout the world.

Synthetic rubbers now account for three fourths of U.S. rubber production, and for one third of all the rubber used in the world. To the list of bunas, neoprene and butyl synthetics devised before World War II, other specialized products have been added. There is a heat- and cold-resistant silicone rubber now being used as a coating on the nozzles of Polaris missiles to prevent the metal from melting at blast-off. The same synthetic is used as insulation of the Polaris afterdome, which houses its guidance equipment. Silicone rubber is also used on spacecraft for gaskets around windows and hatches, and for insulation, because it remains flexible even at low temperatures. Then there is polyurethane, a versatile rubber that may take the form of a soft foam weighing only two pounds per cubic foot, or may be made into an ivory-hard substance with a density of 70 pounds per cubic foot.

Perhaps the ultimate triumph in the search for synthetic rubber is polyisoprene, commercialized in the U.S. in the late 1950s. Polyisoprene *is* natural rubber in every way except that it is made by man. Its advantages and disadvantages are exactly those of natural rubber. Polyisoprene is, oddly enough, almost exactly the same product created by Bouchardat and Tilden in the 19th Century. They failed because they did not have a modern polymerizing catalyst containing aluminum, which creates and controls the "stereo" polymer that is the basis of natural rubber. In a stereo polymer, the small molecules are lined up in a regularly repeating spatial, or three-dimensional, arrangement, like so many cars of the same model, make and type parked bumper-to-bumper on a straight, level road. In a nonstereo polymer, the molecules are in a random pattern as though they were identical beads sewn to a strip of cloth which may be twisted in a dozen different directions.

Stereo catalysts were discovered in 1953 by Karl Ziegler and used on polymers in 1954 by Giulio Natta, both of whom received the Nobel Prize for their contributions. The significance of these achievements extends far beyond a new synthetic rubber, offering greater control over all giant molecules. Science can now build polymers according to plan, for manufacture on a mass basis—a prospect that opens up almost limitless possibilities.

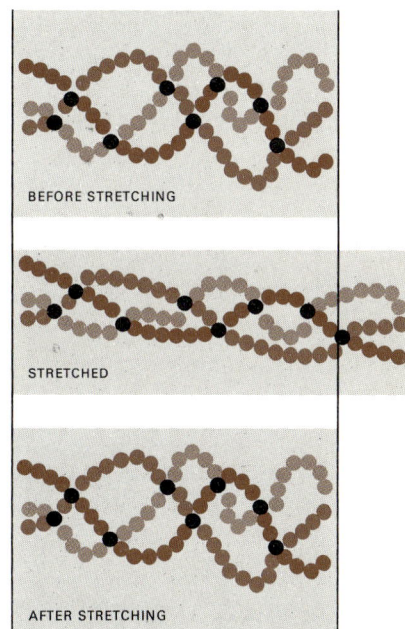

BEFORE STRETCHING

STRETCHED

AFTER STRETCHING

STRETCHED VULCANIZED RUBBER retains its elasticity. The process of vulcanization, which involves heating the rubber with sulfur, produces numerous sulfur cross-links *(black)* between the molecular chains, preventing them from slipping past each other when stretched *(center)*. Released, they spring back to their exact original positions *(bottom)*, and the rubber remains as tough and flexible as it was before.

A Cornucopia
from Oil

Less than a century after the dye industry first mined a treasury of new products from coal, the demands of World War II produced an even more spectacular chemical revolution: the large-scale synthesis of rubber, plastics and other products from hydrocarbon compounds found in oil. These compounds can be obtained from coal, wood, natural gas and grain, but oil is now the primary source: it is cheaper to obtain and process than any other material.

Of the hundreds of petroleum hydrocarbons, or petrochemicals, the most useful are methane, ethylene, propylene, butylene and benzene. These five represent less than three per cent of the output of oil refining processes, but their products account for more than two thirds of the organic chemicals used in the United States. They supply more than half the plastics and fibers, two thirds of the synthetic rubber, two thirds of the soaps and detergents, and appear in almost all cosmetics, pharmaceuticals and insecticides. Nor have the possibilities been exhausted. Tens of thousands of new molecules remain to be made from the five basic hydrocarbons—each a building block for a product yet unknown.

THE PLASTICS INDUSTRY'S BLACK GOLD
Crude oil, here seen bubbling up at the base of an oil well in Alberta, Canada, is the raw material on which most of today's booming plastics industry is based. This great natural resource has been found in every part of the world except Antarctica; although oilmen believe that they have located most of the major concentrations, the search for new crude oil sources still goes on.

The Big Five
of Small Molecules

When crude oil is separated into its components at the refinery, some of the gasoline and naphtha thus obtained are treated to yield five major hydrocarbons: benzene from the gasoline, and methane, ethylene, propylene and butylene from the naphtha. The processes are diagramed at right.

To produce benzene, gasoline is heated. Most of its chain molecules are then turned into rings in the reactor. These are separated by weight until the mixture is converted into high-grade gasoline and benzene.

The naphtha is heated until it has been broken down into smaller molecules, which are separated by weight, liquefied by chilling and compressing, purified by washing and drying, and further separated and chilled, to produce the other four hydrocarbons.

PROCESSING PETROCHEMICALS
The treatment processes by which gasoline and naphtha are converted into the five major hydrocarbons also produce a number of other familiar chemicals. The gasoline yields up hydrogen and gases as well as the high-grade gasolines used as automotive fuel. The naphtha yields up fuel oils for furnaces and diesel engines in addition to lower-grade automotive gasolines.

FURNACE

REACTOR

GASES A
GASOLIN

HYDROGEN

CHILLER

FURNACE

REFINERY

GASOLINE

NAPHTHA

CHILLER AND
COMPRESSOR

SEPARATOR

FUEL OIL

GASES A

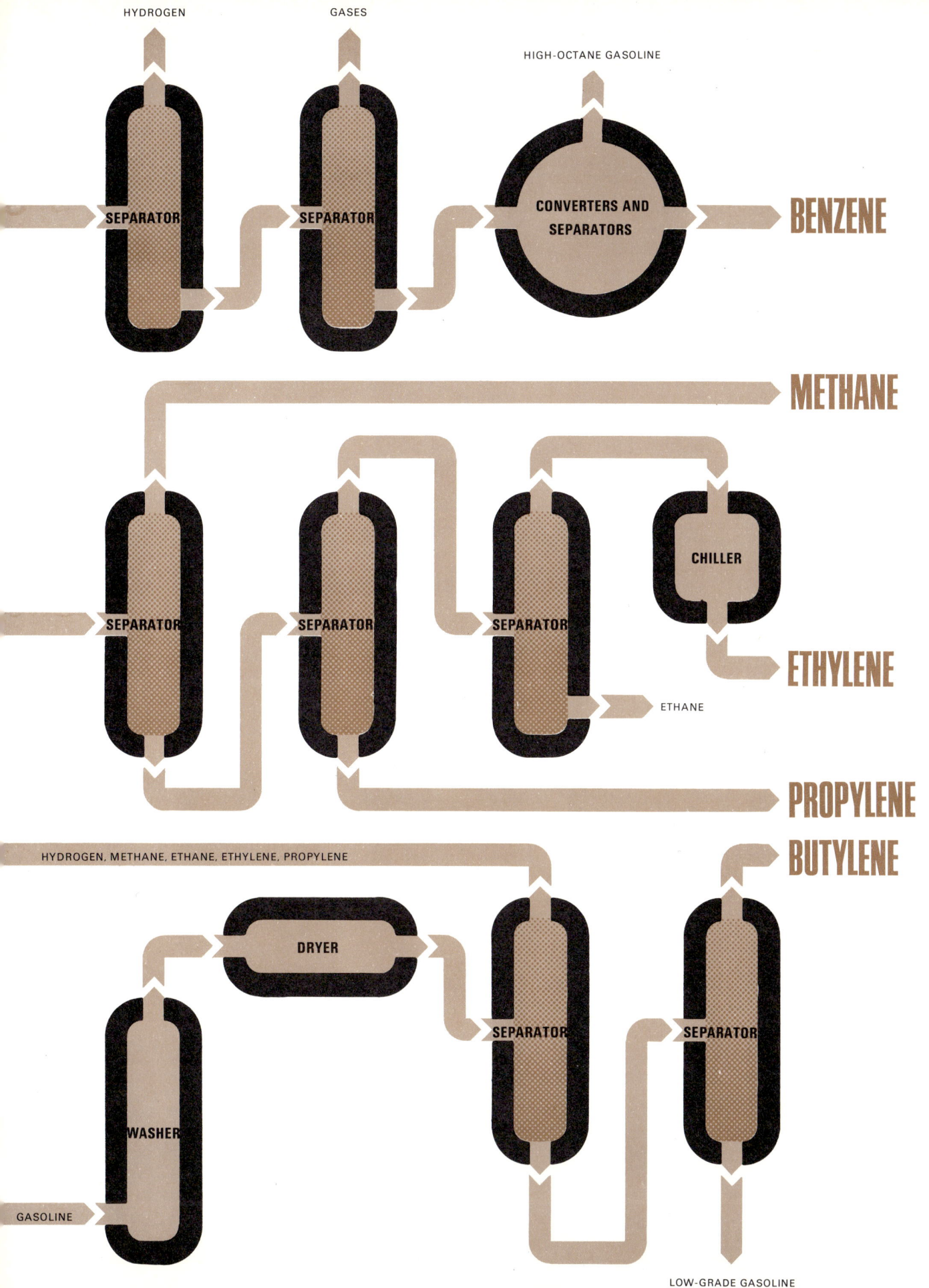

From Cooking Gas to Plexiglas

Methane, whose molecule contains one carbon atom and four atoms of hydrogen, is the base for nearly a hundred products. The processes which transform methane into 18 major products are traced here. Addition of an atom of another element or a molecule of another substance is indicated by a plus sign, subtraction by a minus. Polymerization, which joins small molecules to form giant molecules, is indicated by an "X."

Thus, to transform methane into Plexiglas (sixth product from top in column at right), it is converted into carbon monoxide (CO) by removing its hydrogen atoms and adding one of oxygen. Next, four hydrogen atoms are added to produce methyl alcohol (CH_4O). Then methacrylamide (C_4H_7ON) is added to produce methyl methacrylate ($C_5H_8O_2$). Polymerization transforms this liquid into the glasslike plastic sold under such names as Plexiglas and Lucite.

CH_4
METHANE
COLORLESS, ODORLESS GAS

+ AN ATOM OR MOLECULE IS ADDED.

− AN ATOM OR MOLECULE IS REMOVED.

X THE MATERIAL IS POLYMERIZED.

When more than one reaction occurs in going from one stage to the next, the combination is represented by a grouping of the appropriate symbols.

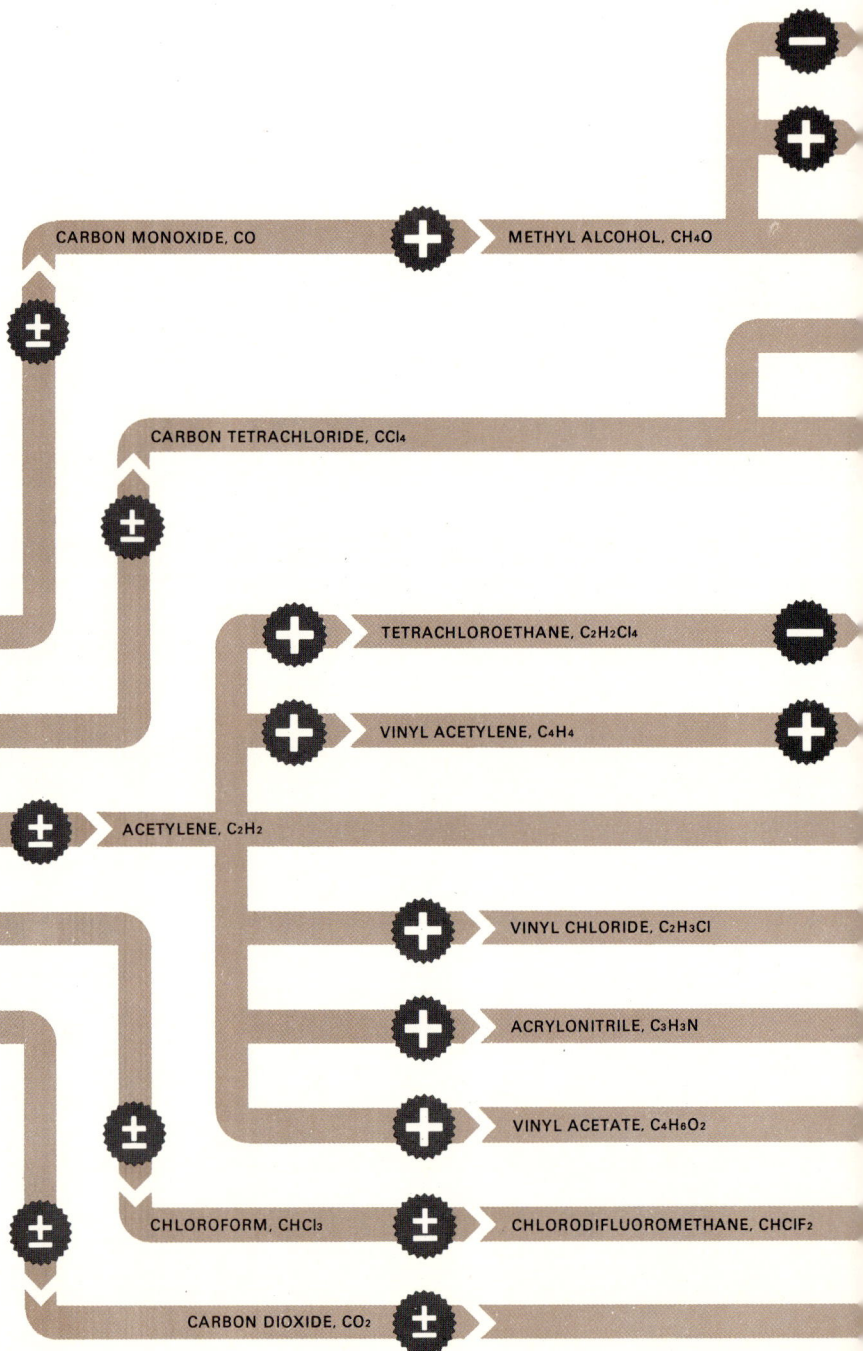

CARBON MONOXIDE, CO

METHYL ALCOHOL, CH_4O

CARBON TETRACHLORIDE, CCl_4

TETRACHLOROETHANE, $C_2H_2Cl_4$

VINYL ACETYLENE, C_4H_4

ACETYLENE, C_2H_2

VINYL CHLORIDE, C_2H_3Cl

ACRYLONITRILE, C_3H_3N

VINYL ACETATE, $C_4H_6O_2$

CHLOROFORM, $CHCl_3$

CHLORODIFLUOROMETHANE, $CHClF_2$

CARBON DIOXIDE, CO_2

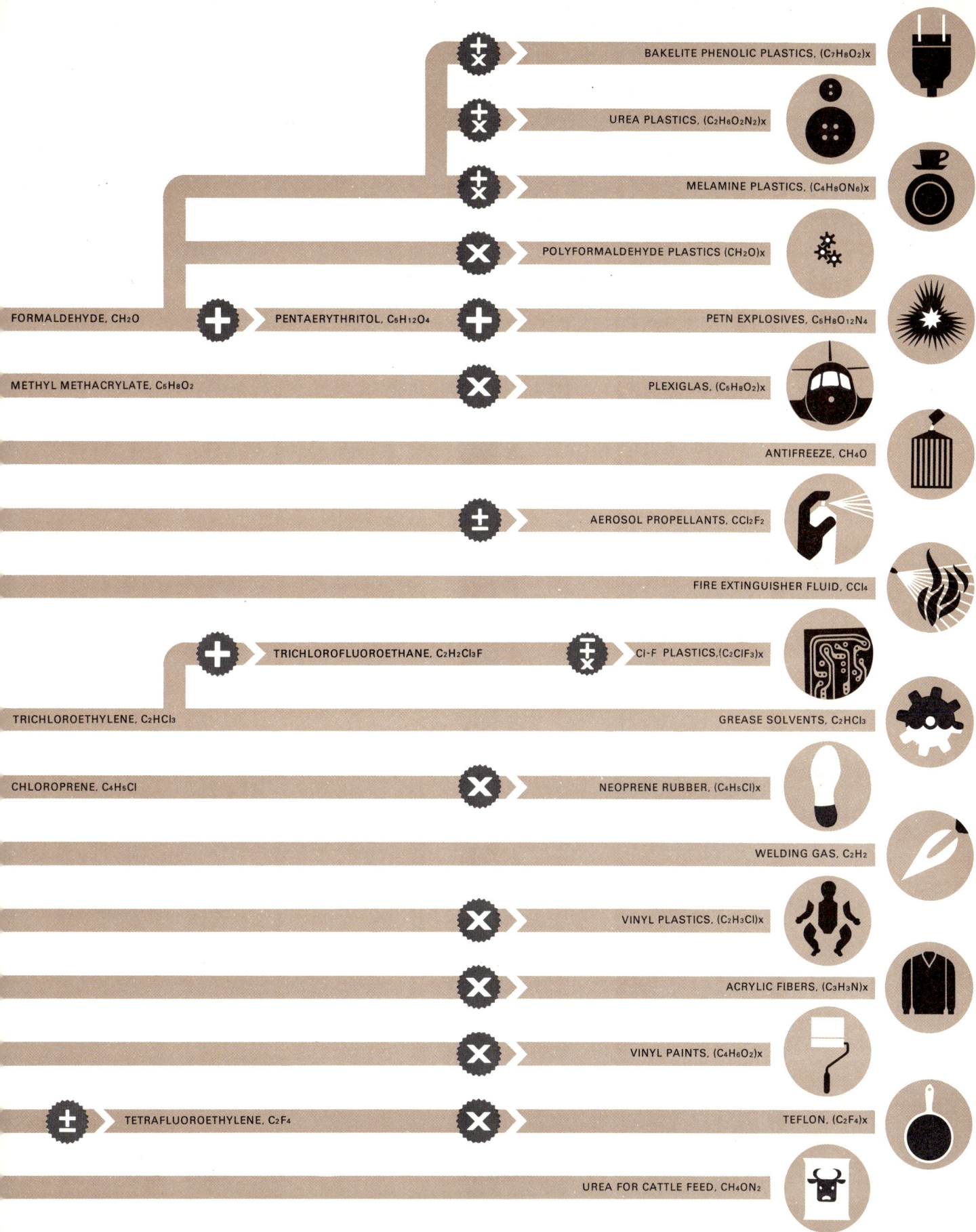

BAKELITE PHENOLIC PLASTICS, $(C_7H_8O_2)x$

UREA PLASTICS, $(C_2H_6O_2N_2)x$

MELAMINE PLASTICS, $(C_4H_8ON_6)x$

POLYFORMALDEHYDE PLASTICS $(CH_2O)x$

FORMALDEHYDE, CH_2O PENTAERYTHRITOL, $C_5H_{12}O_4$ PETN EXPLOSIVES, $C_5H_8O_{12}N_4$

METHYL METHACRYLATE, $C_5H_8O_2$ PLEXIGLAS, $(C_5H_8O_2)x$

ANTIFREEZE, CH_4O

AEROSOL PROPELLANTS, CCl_2F_2

FIRE EXTINGUISHER FLUID, CCl_4

TRICHLOROFLUOROETHANE, $C_2H_2Cl_3F$ Cl-F PLASTICS, $(C_2ClF_3)x$

TRICHLOROETHYLENE, C_2HCl_3 GREASE SOLVENTS, C_2HCl_3

CHLOROPRENE, C_4H_5Cl NEOPRENE RUBBER, $(C_4H_5Cl)x$

WELDING GAS, C_2H_2

VINYL PLASTICS, $(C_2H_3Cl)x$

ACRYLIC FIBERS, $(C_3H_3N)x$

VINYL PAINTS, $(C_4H_6O_2)x$

TETRAFLUOROETHYLENE, C_2F_4 TEFLON, $(C_2F_4)x$

UREA FOR CATTLE FEED, CH_4ON_2

The Busiest Hydrocarbon

For versatility and the quantity of its end products, ethylene, a gas whose molecule has two atoms of carbon and four of hydrogen, is the leader among the basic hydrocarbons. Addition polymerization breaks the double bond between the carbon atoms, converting ethylene into polyethylene *(bottom of chart)*, which accounts for almost 30 per cent of all plastics manufactured in the United States. As a clear film, polyethylene is used to wrap packaged goods, from vegetables to shirts. In heavier form, it is made into containers, and is molded into plastic toys and pipe.

Ethylene can also be made into ethylene glycol, the basis for most car antifreezes; ethyl alcohol, used as a solvent for lacquers and cosmetics, and which also forms the basis for sulfa drugs and explosives; and styrene, used in high-impact plastics, latex paints and most synthetic rubber.

ACETIC ACID, $C_2H_4O_2$

ACETALDEHYDE, C_2H_4O

ETHYL ALCOHOL, C_2H_6O

ETHYLENE GLYCOL, C_2H_6O

ETHYLENE OXIDE, C_2H_4O

ACRYLONITRILE, C_3H_3N

TRIETHANOLAMINE, C_6H_1

C_2H_4 ETHYLENE
COLORLESS, ODORLESS GAS

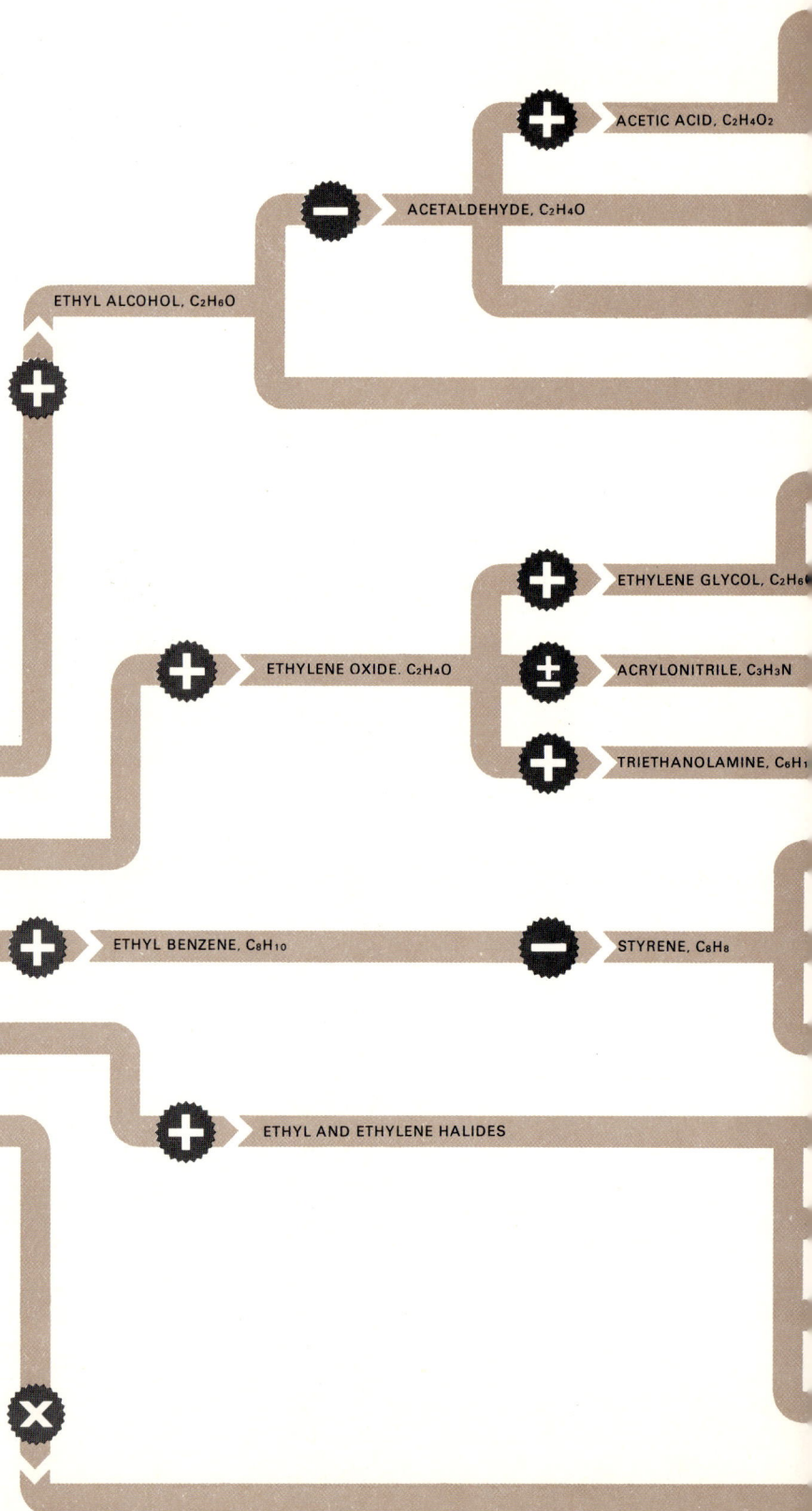

ETHYL BENZENE, C_8H_{10}

STYRENE, C_8H_8

ETHYL AND ETHYLENE HALIDES

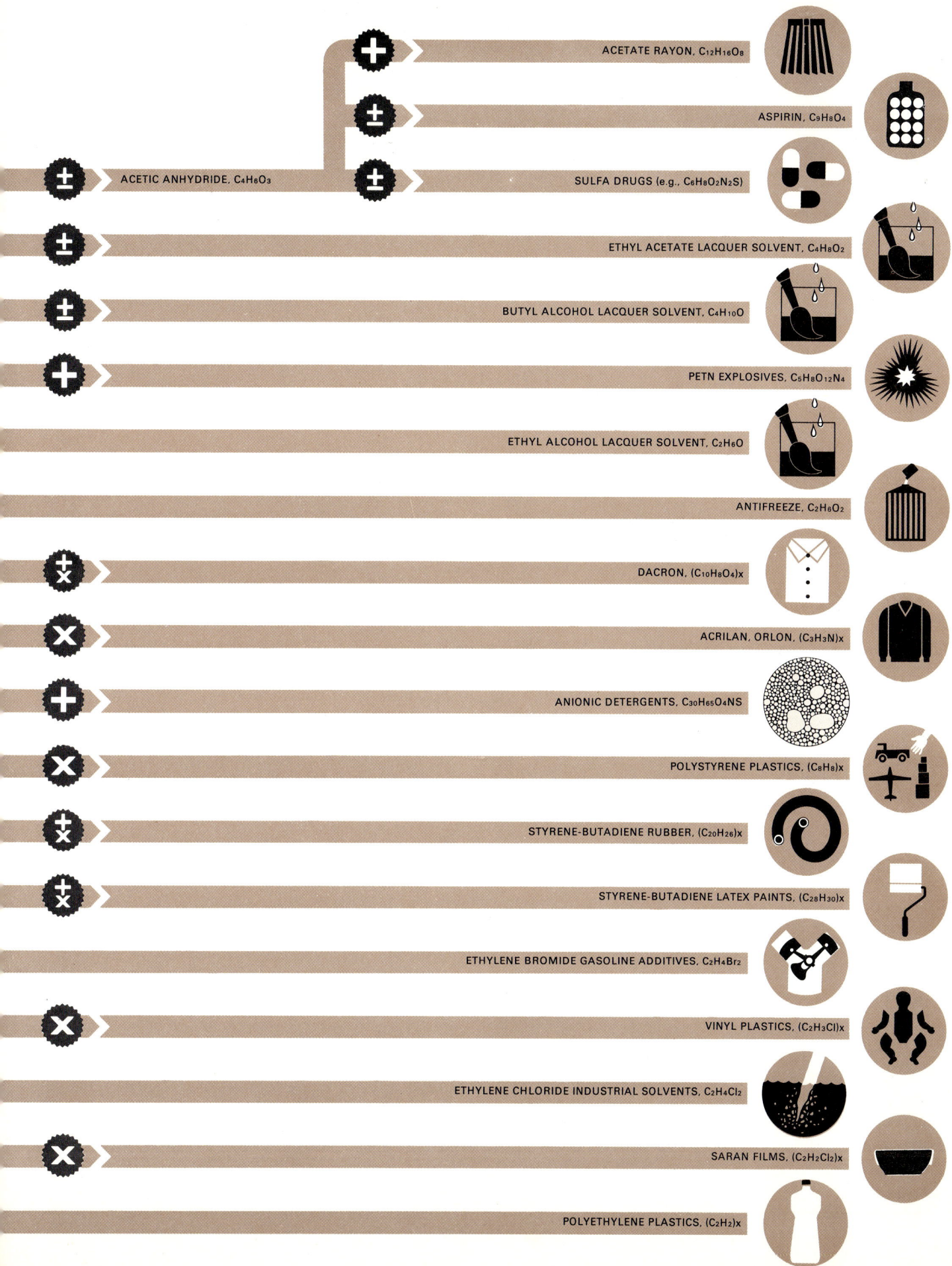

ACETATE RAYON, $C_{12}H_{16}O_8$

ASPIRIN, $C_9H_8O_4$

ACETIC ANHYDRIDE, $C_4H_6O_3$

SULFA DRUGS (e.g., $C_6H_8O_2N_2S$)

ETHYL ACETATE LACQUER SOLVENT, $C_4H_8O_2$

BUTYL ALCOHOL LACQUER SOLVENT, $C_4H_{10}O$

PETN EXPLOSIVES, $C_5H_8O_{12}N_4$

ETHYL ALCOHOL LACQUER SOLVENT, C_2H_6O

ANTIFREEZE, $C_2H_6O_2$

DACRON, $(C_{10}H_8O_4)x$

ACRILAN, ORLON, $(C_3H_3N)x$

ANIONIC DETERGENTS, $C_{30}H_{65}O_4NS$

POLYSTYRENE PLASTICS, $(C_8H_8)x$

STYRENE-BUTADIENE RUBBER, $(C_{20}H_{26})x$

STYRENE-BUTADIENE LATEX PAINTS, $(C_{28}H_{30})x$

ETHYLENE BROMIDE GASOLINE ADDITIVES, $C_2H_4Br_2$

VINYL PLASTICS, $(C_2H_3Cl)x$

ETHYLENE CHLORIDE INDUSTRIAL SOLVENTS, $C_2H_4Cl_2$

SARAN FILMS, $(C_2H_2Cl_2)x$

POLYETHYLENE PLASTICS, $(C_2H_2)x$

A Pair of Gases, Double-bonded

The two hydrocarbons, propylene *(below)*, with three carbon atoms and six of hydrogen, and butylene *(right)*, with four carbons and eight hydrogens, are easy to convert to giant molecules because each has a double bond connecting two of its carbon atoms. Addition polymerization can break this bond and link the small molecules together, converting propylene into polypropylene, its most familiar product, which appears in the forms of fibers, films and molded plastics.

Butylene, which forms the basis for 90 per cent of all synthetic rubbers produced in the U.S., has two forms. The branched-chain form, isobutylene, when polymerized with small amounts of isoprene, forms butyl rubber, used for tires and inner tubes. The straight-chain form, n-butylene, can be converted into butadiene, which combines with acrylonitrile to form oil-resistant nitrile rubbers, and with styrene to form styrene-butadiene rubbers. Combined in different proportions with styrene, butadiene is also a component of latex paints.

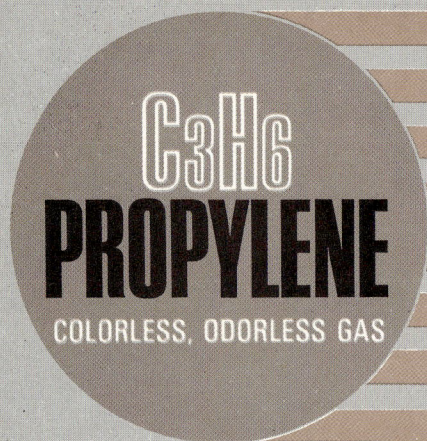

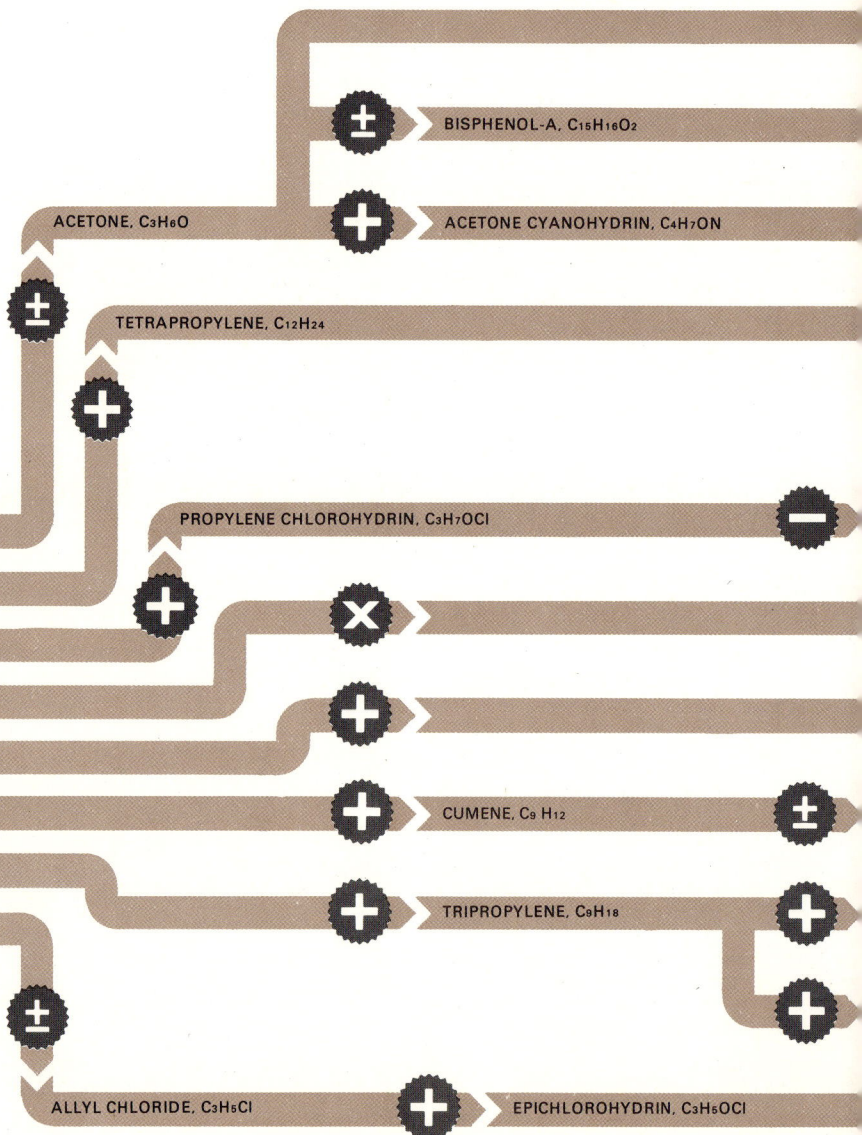

C_4H_8

BUTYLENE

COLORLESS, ODORLESS GAS

BISPHENOL-A, $C_{15}H_{16}O_2$

ACETONE, C_3H_6O

ACETONE CYANOHYDRIN, C_4H_7ON

TETRAPROPYLENE, $C_{12}H_{24}$

PROPYLENE CHLOROHYDRIN, C_3H_7OCl

C_3H_6

PROPYLENE

COLORLESS, ODORLESS GAS

CUMENE, C_9H_{12}

TRIPROPYLENE, C_9H_{18}

ALLYL CHLORIDE, C_3H_5Cl

EPICHLOROHYDRIN, C_3H_5OCl

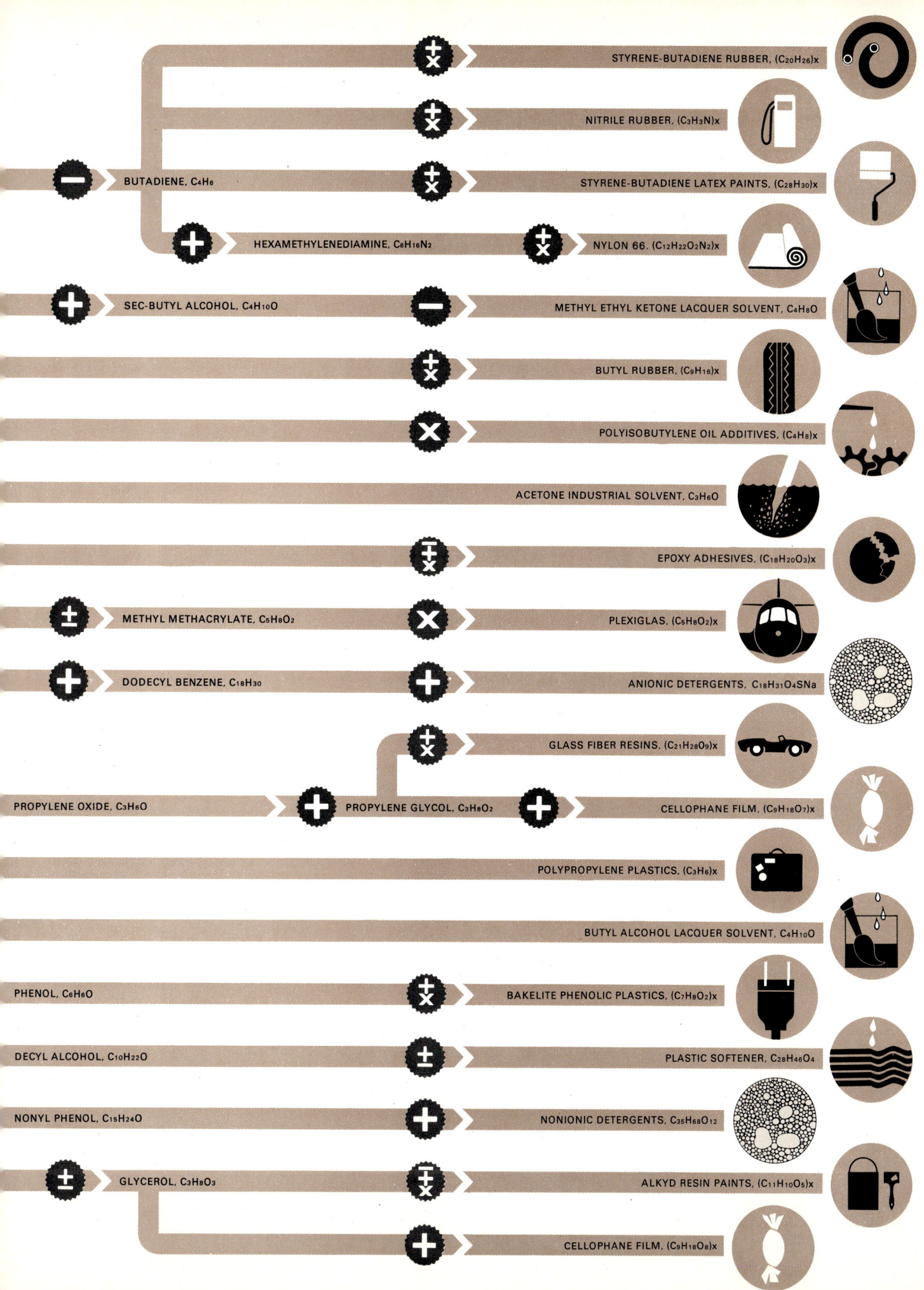

STYRENE-BUTADIENE RUBBER, $(C_{20}H_{26})x$

NITRILE RUBBER, $(C_3H_3N)x$

BUTADIENE, C_4H_6 — STYRENE-BUTADIENE LATEX PAINTS, $(C_{28}H_{30})x$

HEXAMETHYLENEDIAMINE, $C_6H_{16}N_2$ — NYLON 66, $(C_{12}H_{22}O_2N_2)x$

SEC-BUTYL ALCOHOL, $C_4H_{10}O$ — METHYL ETHYL KETONE LACQUER SOLVENT, C_4H_8O

BUTYL RUBBER, $(C_9H_{16})x$

POLYISOBUTYLENE OIL ADDITIVES, $(C_4H_8)x$

ACETONE INDUSTRIAL SOLVENT, C_3H_6O

EPOXY ADHESIVES, $(C_{18}H_{20}O_3)x$

METHYL METHACRYLATE, $C_5H_8O_2$ — PLEXIGLAS, $(C_5H_8O_2)x$

DODECYL BENZENE, $C_{18}H_{30}$ — ANIONIC DETERGENTS, $C_{18}H_{31}O_4SNa$

GLASS FIBER RESINS, $(C_{21}H_{28}O_9)x$

PROPYLENE OXIDE, C_3H_6O — PROPYLENE GLYCOL, $C_3H_8O_2$ — CELLOPHANE FILM, $(C_9H_{18}O_7)x$

POLYPROPYLENE PLASTICS, $(C_3H_6)x$

BUTYL ALCOHOL LACQUER SOLVENT, $C_4H_{10}O$

PHENOL, C_6H_6O — BAKELITE PHENOLIC PLASTICS, $(C_7H_8O_2)x$

DECYL ALCOHOL, $C_{10}H_{22}O$ — PLASTIC SOFTENER, $C_{28}H_{46}O_4$

NONYL PHENOL, $C_{15}H_{24}O$ — NONIONIC DETERGENTS, $C_{35}H_{68}O_{12}$

GLYCEROL, $C_3H_8O_3$ — ALKYD RESIN PAINTS, $(C_{11}H_{10}O_5)x$

CELLOPHANE FILM, $(C_9H_{18}O_6)x$

A Liquid and Its Flow of Products

Two characteristics lend benzene its distinction among the five basic hydrocarbons. It is a liquid rather than a gas, and its six carbon and six hydrogen atoms are arranged in a ring rather than a chain. The ring structure makes benzene (not to be confused with the common solvent, benzine) an invaluable base for products as varied as aspirin and insecticide.

When heat or catalysts are used to open the ring and convert it into a chain, the characteristics of the molecule change, making it useful in the production of such materials as nylon and polyurethane foam, a synthetic substitute for foam rubber.

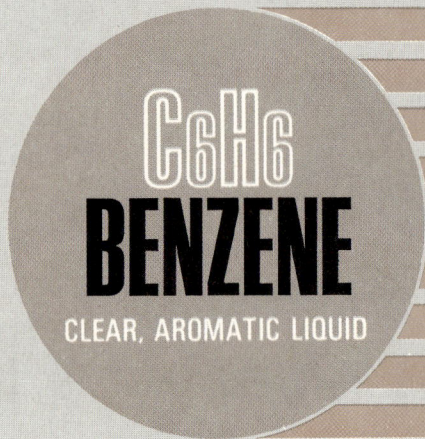

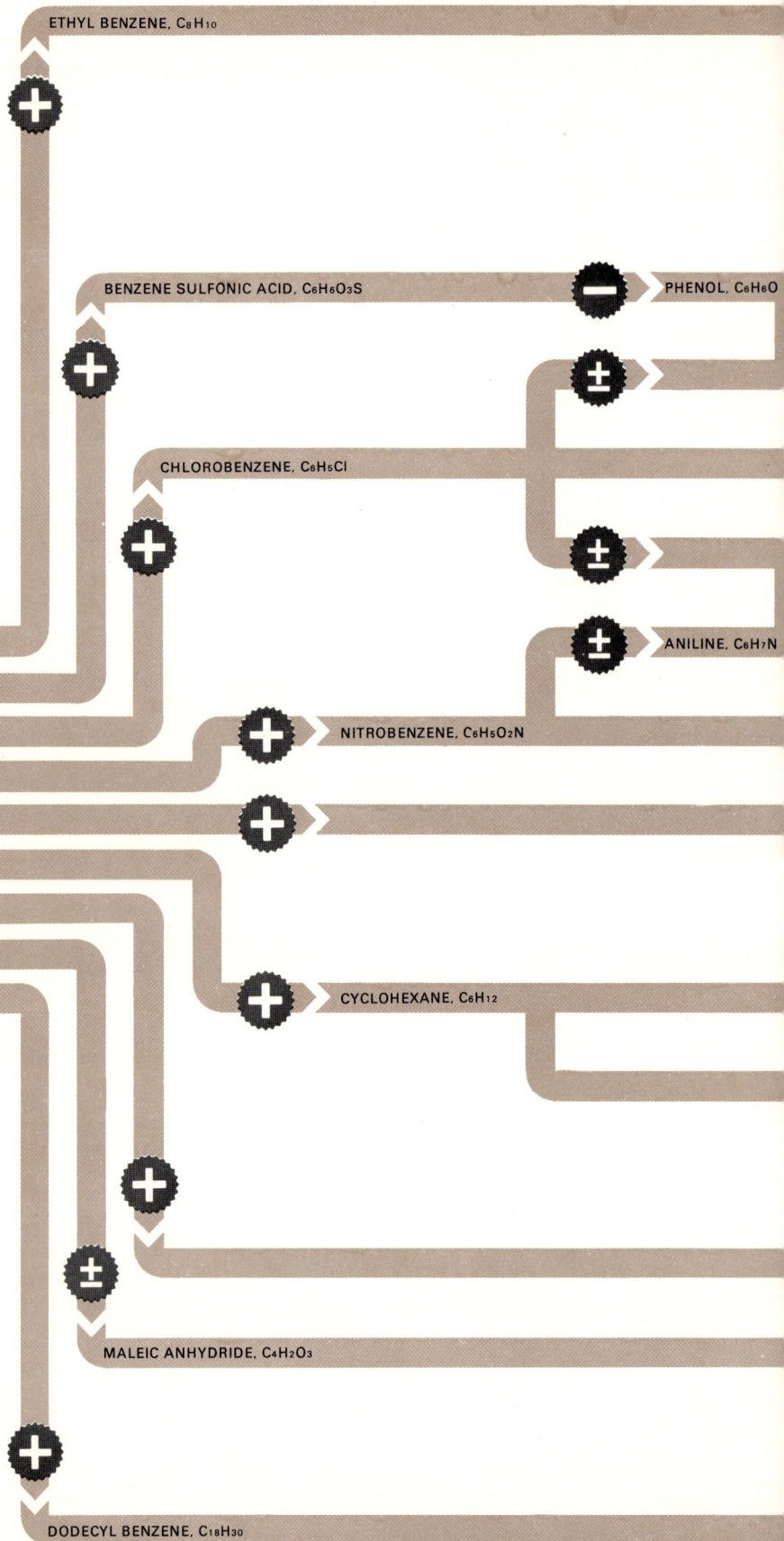

C_6H_6

BENZENE

CLEAR, AROMATIC LIQUID

ETHYL BENZENE, C_8H_{10}

BENZENE SULFONIC ACID, $C_6H_6O_3S$

PHENOL, C_6H_6O

CHLOROBENZENE, C_6H_5Cl

ANILINE, C_6H_7N

NITROBENZENE, $C_6H_5O_2N$

CYCLOHEXANE, C_6H_{12}

MALEIC ANHYDRIDE, $C_4H_2O_3$

DODECYL BENZENE, $C_{18}H_{30}$

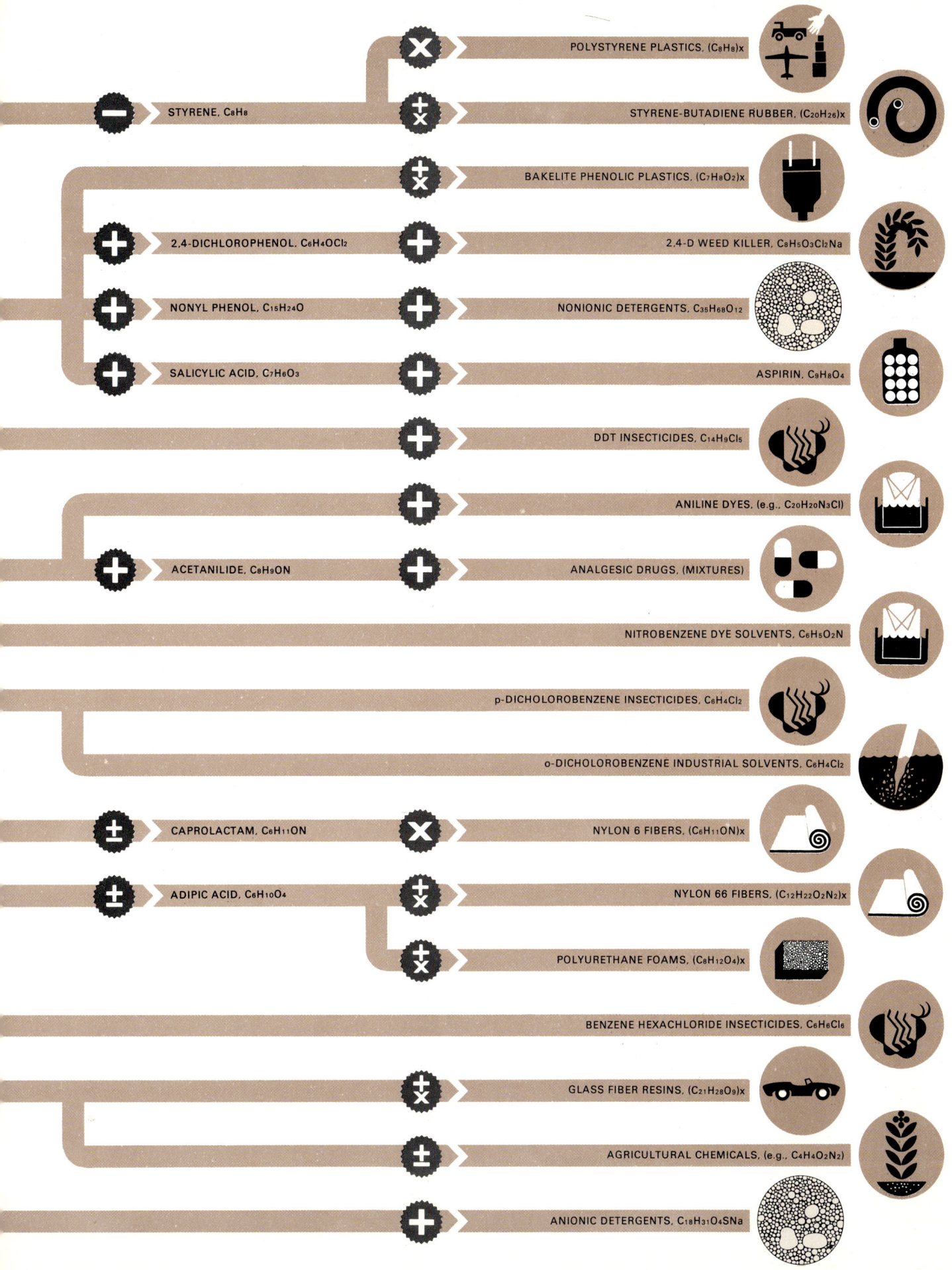

POLYSTYRENE PLASTICS, $(C_8H_8)_x$

STYRENE, C_8H_8

STYRENE-BUTADIENE RUBBER, $(C_{20}H_{26})_x$

BAKELITE PHENOLIC PLASTICS, $(C_7H_8O_2)_x$

2,4-DICHLOROPHENOL, $C_6H_4OCl_2$

2,4-D WEED KILLER, $C_8H_5O_3Cl_2Na$

NONYL PHENOL, $C_{15}H_{24}O$

NONIONIC DETERGENTS, $C_{35}H_{68}O_{12}$

SALICYLIC ACID, $C_7H_6O_3$

ASPIRIN, $C_9H_8O_4$

DDT INSECTICIDES, $C_{14}H_9Cl_5$

ANILINE DYES, (e.g., $C_{20}H_{20}N_3Cl$)

ACETANILIDE, C_8H_9ON

ANALGESIC DRUGS, (MIXTURES)

NITROBENZENE DYE SOLVENTS, $C_6H_5O_2N$

p-DICHLOROBENZENE INSECTICIDES, $C_6H_4Cl_2$

o-DICHLOROBENZENE INDUSTRIAL SOLVENTS, $C_6H_4Cl_2$

CAPROLACTAM, $C_6H_{11}ON$

NYLON 6 FIBERS, $(C_6H_{11}ON)_x$

ADIPIC ACID, $C_6H_{10}O_4$

NYLON 66 FIBERS, $(C_{12}H_{22}O_2N_2)_x$

POLYURETHANE FOAMS, $(C_8H_{12}O_4)_x$

BENZENE HEXACHLORIDE INSECTICIDES, $C_6H_6Cl_6$

GLASS FIBER RESINS, $(C_{21}H_{28}O_9)_x$

AGRICULTURAL CHEMICALS, (e.g., $C_4H_4O_2N_2$)

ANIONIC DETERGENTS, $C_{18}H_{31}O_4SNa$

7
Molecules Made to Order

A skater glides across an experimental surface made of a slippery plastic called Teflon.

THE MODERN POLYMER CHEMIST is a designer-builder of giant molecules. No longer a blind experimenter or pure theoretician, he has now acquired such mastery over his materials that he can outline in advance the properties of the substance he wants to make, and then proceed to make it. His intimate understanding of the structure and properties of small molecules and the techniques of manipulating them enables him to combine them into giant molecules more or less to order.

He therefore enjoys an enormous advantage over even such recent predecessors as Baekeland and Carothers. Although neither of these men made his great contribution to science by accident—Baekeland set out to create a synthetic resin, and Carothers' goal was the fabrication of a synthetic fiber—neither scientist could know what properties the finished product would have until the work was successfully completed.

Today the polymer chemist, like the architect, incorporates the needed characteristics of his creation in the design. Without this capability of making predesigned giant molecules, much of contemporary technology would be impossible. Everything from the insulation to protect a space capsule against the searing heat of reentering the earth's atmosphere to a safe substitute for a damaged heart artery has come from the laboratory of the polymer chemist, who deliberately created giant molecules to do the specific jobs.

Among these new synthetic materials are several that stand out both for their utility and for the ingenuity of their design and manufacture. The fluorocarbon polymers, spandex, the ABS plastics and the group of materials called fibrids are totally different from one another and from any natural materials. One is the opposite of an adhesive—nothing at all will stick to it; one combines the opposing qualities of a fiber and a rubber; the third is as tough as metal but far less heavy; and the fourth adapts some of the most unusual properties of polymers to equally unusual uses.

Of all the new materials, the fluorocarbon polymers are among the most versatile and valuable. Until they were developed, no giant molecule could withstand the effects of both oil and water; if it repelled the one, it accepted the other. Raincoat material might resist a downpour, but it soaked up the oils of a spilled ice-cream cone. Nylon shrugs off oil but nylon fabrics can be made wet. The fluorocarbon polymers are equally unaffected by both oil and water. All liquids roll off them.

The fluorocarbons, like most giant molecules, are built on a backbone of carbon atoms. But their remarkable properties come from fluorine, one of the strangest elements in the chemist's armamentarium. As an element, fluorine is a pungent, poisonous, yellow-green gas. In contrast with carbon, which passively accepts partnership with other atoms, fluorine is extremely reactive, and is never found alone in nature but only combined in inorganic molecules. It reacts so strongly with silicon that it eats away silicon-containing rocks and glass. It combines easily with hydrogen and an instant's contact sets fire to wood and wool. Hydrogen fluoride, a gaseous compound of hydrogen and fluorine, is so dangerous to human tissue that many early experimenters who worked with it lost parts of their hands.

Yet, precisely because fluorine is so reactive, the compounds that contain it are extremely inert. Once the fluorine atom attaches itself to the atom of another element, it holds fast, refusing to combine with the atoms of other substances.

Fluorine itself was first isolated in 1886 by Henri Moissan, the French chemist and Nobel laureate. And Moissan also produced the first compound of carbon and fluorine, carbon tetrafluoride, a gas which in structure is analogous to the very simple molecule of methane, an ordinary cooking gas, composed of four atoms of hydrogen bonded to an atom of carbon. As could have been anticipated by fluorine's reactivity, carbon tetrafluoride is a very different kind of gas from methane. Methane readily combines with oxygen to burn. Carbon tetrafluoride, because of its inertness, puts out fires.

It was primarily this inertness that led to further research with the fluorocarbon gases. Chemists knew that they would be harmless—they would not catch fire, corrode containers or endanger people who accidentally inhaled them. For these reasons, they should prove valuable as refrigerant gases in the coils of refrigerators, and as propellants in spray cans of such liquids as insecticides, hair spray and shaving cream. By the 1930s, a number of fluorocarbon gases were in commercial manufacture for just these purposes. But in 1938, by one of those happy accidents in which the history of science abounds, the giant fluorocarbon molecules were brought into being.

Mystery of the empty tank

At that time, a research team was performing some experiments with one of the Freon gases, a newly developed family of fluorocarbon refrigerants that are still in use. In the course of the work, Du Pont chemist Roy J. Plunkett filled several small cylinders with tetrafluoroethylene, a fluorocarbon gas whose molecule is composed of four atoms of fluorine and two of carbon. When the cylinders were filled, he sealed their valves and stored them overnight.

The next morning, he noticed that the pressure gauge on one of the cylinders had dropped to zero, indicating that it was empty. But when he examined the tank he found that the seal was still tight and, putting it on the scale, he discovered that it weighed as much as when it was full. He opened it, turned it upside down and shook it. A waxy, white powder fell out.

It was immediately obvious what had happened. The tetrafluoroethylene had polymerized spontaneously, transforming its small molecules into solid giant molecules. Precisely why the polymerization had occurred

MAKING POLYETHYLENE *(far right)*, the familiar white plastic used for containers, chemists convert small gaseous ethylene molecules *(right)* into a solid. The same ethylene molecules can be polymerized into two types of polyethylene, one a soft material and the other a hard substance *(shown here)*. This type of polyethylene is prepared by using catalysts to force the ethylene's double carbon bonds to open, and the molecules to link together in a long, single-bonded polyethylene chain.

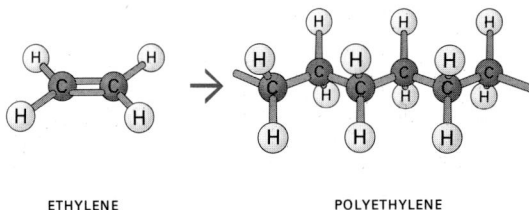

ETHYLENE   POLYETHYLENE

in that particular cylinder, chemists still do not know. But the event itself was not so surprising. Only a few years earlier, R. O. Gibson and E. W. Fawcett of Great Britain's Imperial Chemical Industries Limited had created polyethylene, the giant molecule of squeeze-bottle plastic, by a similar accident. And ethylene, the petroleum derivative from which polyethylene is built, is the hydrocarbon counterpart of Plunkett's tetra-fluoroethylene.

When tetrafluoroethylene polymerizes, a break occurs in the double bond that joins the two carbon atoms of the molecule. (This is the only reactive link in the structure; double bonds are always easier to activate than single bonds.) One of these bonds then shifts around to connect neighboring molecules to one another. Through this simple process, a group of small molecules, each containing only six atoms, is transformed into a giant molecule—a long, regular chain composed of tens of thousands of interconnected tetrafluoroethylene units, best known under the trade name of Teflon.

Teflon is more inert and stable than the hydrocarbon, polyethylene, or the fluorocarbon gas from which Teflon is derived. It is more inert and stable than tetrafluoroethylene because it contains only single bonds. And it is more inert and stable than polyethylene because the fluorine atoms, being somewhat larger than those of hydrogen, surround the carbon atoms and shield them almost perfectly, refusing to be jogged loose by any chemical forces. Teflon is, indeed, chemically more inert and stable than any other natural or man-made resin. The fluorine atoms are so tightly bonded to the carbon atoms and the carbon atoms to one another that it is extremely difficult to break their connections and thus permit them to link up with the atoms of other substances. Because of this aloofness, Teflon will not burn, corrode or otherwise damage any material with which it comes in contact.

The impregnable plastic

Teflon will not pass electric currents. Because it will not combine with oxygen, Teflon cannot sustain life. It is therefore immune to attack by fungi, molds or pests. No liquid has as yet been found that will dissolve it. Because it will not combine with acids, it cannot be corroded: a bath in boiling nitric and sulfuric acids only cleanses it of dirt. Nor do temperature changes affect the Teflon molecule. Teflon does not begin to melt until 620° F., 200 degrees higher than the melting point of tin. And even then it does not become a liquid, but a translucent gel. It is equally unaffected by cold, remaining unchanged at temperatures as low as −450° F., only a few degrees above absolute zero. Because Teflon resists forming even temporary bonds with other molecules, nothing sticks to it. Other substances slide over it so readily that fabrics made of Teflon

A MOLECULE OF TEFLON is similar to the polyethylene molecule *(opposite)*, except that fluorine has replaced the hydrogen atoms on its carbon chain. Teflon is polymerized from tetrafluoroethylene, whose double bonds are broken to form the giant molecule. The plastic has gained great popularity as a frying-pan coating because food will not stick to it. This property results from the fact that fluorine atoms, once bonded to carbon, are so unreactive that they repel almost all other substances.

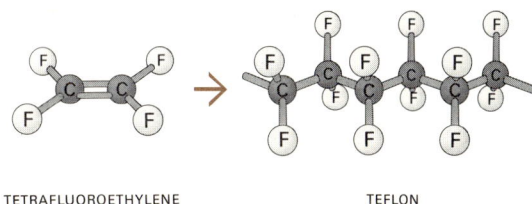

TETRAFLUOROETHYLENE   TEFLON

have been used—for demonstration purposes only—in place of ice on skating rinks.

Teflon found its first uses in the Second World War. Shortly after its discovery, it was pressed into service in the Manhattan Project, which produced the atomic bomb. Engineers were faced with the problem of finding a gasketing material inert enough to withstand the corrosive effects of the fluorine gas used to separate the fissionable isotope of uranium 235 from its inert counterpart, uranium 238. Teflon proved to be the answer, as it proved to be the answer to many other military needs. During the War it was also used to insulate the wiring of radar equipment and airplane motors, and to line tanks of liquid fuel so cold that linings made of conventional materials froze.

Adapting Teflon to peace

With the end of the War, experimentation began to seek ways of putting Teflon's extraordinary properties to wide peacetime use. It was not an easy task. The very qualities of heat resistance and insolubility that made the plastic so valuable also made it extremely difficult to work with. Unless it could be modified and new methods could be found of handling it, Teflon would remain nothing more than an expensive strategic material.

It was at this point that the polymer chemist's understanding of molecular structure and design paid off. It was known, for example, that many of Teflon's properties were related to the regular chain structure of its giant molecule. If the chain were made less regular, the plastic would still be strong and inert, but it would be more manageable. It would have a somewhat lower melting point and would show less tendency to crack when it hardened. To produce this less regular chain, chemists designed a copolymer, combining tetrafluoroethylene with another fluorocarbon gas. They copolymerized tetrafluoroethylene with hexafluoropropylene, whose molecule is composed of six atoms of fluorine and three of carbon arranged in a more irregular pattern than that of tetrafluoroethylene. The resulting giant molecule, called Teflon FEP, is finding increasing use in the engines of automobiles, jet planes and rockets, where its resistance to oils and heat gives it advantages over any conventional material in gaskets, insulators and shock absorbers.

Other techniques of molecular design uncovered additional ways of modifying the original molecule to put it to wider use. By creating a fluorocarbon-containing molecule derived from a fluorinated acid, chemists have produced a liquid coating which, when applied to paper and fabrics, keeps them free of oil and water stains. The acid end of the hybrid fluorocarbon molecule reacts with the paper and fiber molecules and therefore sticks to them. But because the fluorocarbon end of the hybrid does not react with other molecules, it resists combining either with the paper or fiber, or with oily or watery liquids. Thus the hybrid molecule provides a sticky side, which adheres to the material it is protecting, and a nonsticky side, which wards off oil and water. Marketed under such trade names as Scotchgard and Zepel, this substance forms a stain-proof coating on upholstery fabrics, raincoats and slipcovers.

High-flying rubber

Fluorocarbon rubber, still another modification of the basic polymer, is used to coat the wings of aircraft and keep them ice-free even at high altitudes. It is created by altering the composition of the small molecules from which the polymer is built. Instead of substituting fluorine atoms for all of the atoms of hydrogen in the original hydrocarbon molecule, chemists replace only some, to produce a molecule containing hydrogen, carbon and fluorine. When such small molecules are polymerized, either alone or with pure fluorocarbon molecules, a fluorocarbon rubber is the result. In a somewhat similar way, fluorocarbon film has been developed, which, when applied as a coating, keeps house sidings free from the effects of weather.

Finally, a way was found to produce fine dispersions of Teflon by suspending it in water. This makes it possible to spray the plastic on various surfaces. Teflon had already been put to use in industry. As a lining for tanks, it saved these containers from corrosion by acids. In commercial kitchens, it was used to coat the conveyor belts that carry candies and baked goods. The discovery of the dispersion technique made it economically feasible to put Teflon to consumer use. As most housewives know, this plastic is an ideal nonstick coating for pots and pans, and for steam irons as well. Increasingly, it is being put to use as coating for oven panels, which makes them easier to clean.

As the fluorocarbon plastics combine two properties previously considered incompatible in giant molecules—the properties of repelling both oil and water—so the man-made giant molecule of spandex combines two other properties that were previously considered mutually exclusive: fibrous strength and rubbery elasticity. Spandex is a strong and elastic thread. In orthodox chemical thought, the very notion of a rubbery fiber was a contradiction in terms.

The reason is simple. Rubber's great ability to stretch is conferred on it by the flaccidity of the giant molecules of which it is composed. The rubber molecule, as has been said, is like a piece of cooked spaghetti, and a piece of rubber is like a dish of spaghetti: a group of flaccid strings all tangled together, touching one another at various random points so that they are easily pulled apart. On the other hand, a fiber's strength and toughness are given it by the relative rigidity of its giant

THE SPANDEX MOLECULE is unique among synthetics because it combines an elastic carbon-hydrogen-oxygen component, part of which has been illustrated at lower left, with a rigid fiber component of linked benzene rings (directly below) to produce a "stretch" fabric ideal for bathing suits, ski pants and women's undergarments. Experts had never thought that such combinations were possible until Du Pont chemist William H. Charch succeeded in making the first spandex, marketed in 1960.

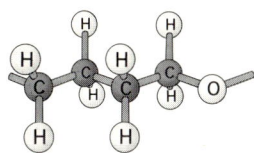

ELASTIC COMPONENT   FIBER COMPONENT

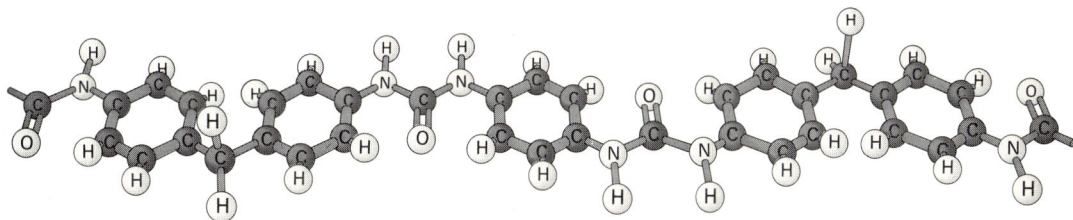

molecules and by the strong forces that hold neighboring molecules together. To carry the spaghetti image a step further, a fiber thread is like uncooked but fused spaghetti strands—a group of long stiff stalks that lie parallel and are firmly fastened to one another. To bring such differing structures and qualities together in one molecule demanded an enormously complex piece of molecular designing.

The prolific chemist

The man who performed this feat was William Hale Charch, Director of Pioneering Research at Du Pont's Textile Fibers Department from 1935 until his death in 1958. Charch held 53 patents in the polymer field. As Herman Mark wrote, shortly after Charch's death: "Only very rarely in modern inventorship has one single man initiated so many . . . products. . . . It is necessary to quote such illustrious names as Edison, Langmuir and Kettering to reach the same level of ingenuity, resourcefulness and perseverance." But with all his achievements and the honors that they brought, Charch remained a chemist's chemist, who stayed close to the creative atmosphere of the laboratory even after he became an administrator, and whose enthusiasm infected and inspired everyone who worked with him.

The starting point for Charch's development of spandex fiber was the search for a rubberlike material whose molecule was strong enough to withstand the rough treatment necessary to draw it into a fine thread. No conventional rubber—natural or synthetic—had such strength and resistance: the thin strands of rubber used in elastic, for example, were cut from sheets, not spun or extruded into threads. However, two somewhat unconventional synthetic rubbers had been produced that seemed to offer hope for a rubber fiber. One was Vulcollan, based on a compound of glycol and adipic acid, which had been developed in the 1930s by the German chemist Otto Bayer. The other was Adiprene, a polyether rubber manufactured by Du Pont. But even these materials, although stronger than any others available, could not be used unmodified. The individual molecules were not sufficiently rigid, nor were adjacent molecules held together by an adequate number of intermolecular bonds.

There was, however, one possible remedy for both deficiencies: to incorporate into these rubbery molecules a series of fibrous—i.e., non-rubbery—segments which would supply both the necessary rigidity and the anchor points for additional intermolecular bonds. Fortunately, as Charch knew, the linking points for such segments were already built into both Vulcollan and Adiprene. The molecules of both materials consisted of a series of relatively short rubbery blocks linked together by so-called diisocyanate bridges: compounds containing two groups of carbon, nitrogen and oxygen atoms, each of which was connected to an end of each rubbery segment. It was the introduction of these bridges that had made both Vulcollan and Adiprene possible in the first place. Not only did the bridges link a series of relatively short blocks into a giant molecule, they also made it possible to bond together the polymers thus created. By opening the diisocyanate bridges and inserting a nylonlike segment into each opening, Charch solved all his problems simultaneously. He increased the length of the giant molecule; he incorporated a rigid segment into it; and he provided, in that segment, additional bonds between the molecules.

Once he had arrived at the broad outline of his rubbery fiber, Charch was faced with the problem of determining the details of its molecular structure, so that it would have precisely the properties that he wanted. What should be the overall pattern in which the segments were arranged? It was evident that the rubbery segment should be the longer one. Only such a structure could provide the necessary resilience. But the proper balance between strength and resilience depended on the way in which the rubbery and fibrous segments were arranged. Spandex would have to be a so-called block copolymer, in which the differing segments fell in a regular alternating pattern. If its two types of segments were merely to fall in a random order, the material might not have the necessary balance of properties.

A problem in production

Charch devoted more than five years to his work on spandex, and it was his final contribution to polymer chemistry. He died only a few months after its development was announced. Nor could Fiber K, as it was then called, be put into immediate commercial manufacture. Like so many other man-made materials during the initial stages after their development, Fiber K had every desirable property save one: it could not be mass-produced with the manufacturing tools and techniques then available. It took two years to discover ways of mass-producing spandex. But in the early 1960s, spandex fibers began to appear on the market in a wide variety of guises. They are marketed under different trade names, which reflect somewhat different chemical compositions. Du Pont's Lycra is based on a polyether similar to Adiprene. Uniroyal's Vyrene, on the other hand, has its origins in a polyester, more like Vulcollan. But whatever their chemical bases, all spandex fibers have essentially the same structure and, therefore, the same properties. And they have found widespread use in bathing suits, ski suits, upholstery materials and girdles.

In developing spandex, Charch mated seemingly incongruous properties by combining two molecules that separately possessed the traits he sought. This game can be played with more than two starting materials. One of the most useful of the new plastics, ABS, is a combination of three molecules: acrylonitrile, butadiene and styrene.

This remarkable material, the first giant molecule to mimic successfully the most useful properties of metal, was initially developed in response to the demands of World War II. During the early days of the conflict, the American armed forces urgently needed a material to be used as housing for such aircraft equipment as radar and fuel tanks. To serve this purpose, it had to be lightweight, rigid and shatterproof.

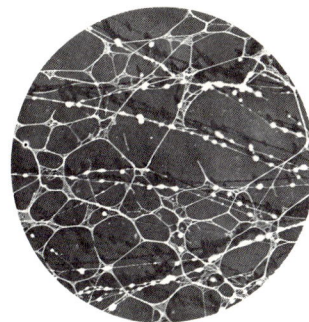

A NONWOVEN FABRIC, the synthetic material shown above, greatly magnified, can be made from many polymers, including nylon, with a new process called fibridization. In one method, the polymer in solution is forced through an extruder into an agitated bath. The polymer hardens in the bath, and the action splinters the strands, creating a network pattern called a fibrid. The fabric, which resembles filter paper, is used for electrical insulation and other purposes.

The task of developing the new substance fell to a joint team of industrial and research chemists headed by Lawrence Daly, a young chemical engineer at Uniroyal. Daly and his colleagues undertook the task with a certain number of misgivings. Though many materials are shatterproof, and many are rigid, the two properties had never before been combined. In conventional terms, they were as incompatible as the properties of rubber and fiber that Church brought together. But by copolymerizing a resinous compound of acrylonitrile and styrene that was both rigid and hard with a tough, shatterproof polymer of butadiene and acrylonitrile, Daly's team created precisely the material that was needed; by 1942 it was in use.

A gift to golfers

Its first peacetime application, in 1946, was far removed from its original purpose. As soon as ABS was put into commercial production, it was ordered in quantity by the manufacturers of golfing equipment, who recognized that it would solve one of their most vexing problems: that of finding a material sturdy enough to serve as a covering for the bottoms of golf bags. But ABS was not long confined to this use. In a few brief years its application had spread to include such varying products as luggage, water pipe, and radiator grilles for automobiles. Like other plastics, it is inexpensive, light, corrosion-resistant, and easy to form into a variety of shapes and sizes. Like metal, it is both rigid and tough: it holds its shape precisely and sustains heavy loads, yet it also stands up to sudden blows. Many plastics are rigid but brittle and others are tough but rubbery; the ABS family of plastics is the first to achieve rigidity without brittleness, toughness without rubberiness.

This unique combination of properties might be suspected from characteristics associated with the starting materials. The styrene in ABS is a liquid; when it is polymerized it becomes the familiar plastic of wall tiles, radio housings and combs—rigid but brittle. Acrylonitrile, a colorless, faintly pungent liquid, also polymerizes very usefully, to form the base for the woolly fibers of Acrilan and Creslan which are used in blankets, sweaters and carpets. Butadiene, a gas derived from petroleum, polymerizes alone or in combination with either styrene or acrylonitrile into a very good synthetic rubber. None of these man-made molecules, however, approaches a metal. They are fibrous, or rubbery, or hard and brittle, but they are not rigid and tough.

The metallic property appears only when all three simple molecules are linked in the ABS molecule. Most ABS plastics contain approximately 20 per cent acrylonitrile, 20 per cent butadiene and 60 per cent styrene. By varying these proportions somewhat and by varying methods of preparation and manufacture, materials can be produced with properties quite different from one another. For use in dishwashers, for example, a plastic is needed that is unaffected by heat, water and constant hard wear. Here, high levels of acrylonitrile and styrene must be used, as they are in making pipe. On the other hand, plastics to be made into refrigerator parts must have considerable chemical resistance and must be able to withstand such repeated impacts as the slamming of doors. To produce this property, a high level of butadiene is required. Casings for telephone sets must be easily moldable. To produce such a material, a particular technique, graft polymerization, must be used. This process involves adding side chains of one polymer to the linear chain of another, to create a more flexible substance. And other variations in technique produce still other variations in properties, with the result that the ABS plastics are valuable not only for these heavy uses but also for the heels of women's shoes and for toys.

Beyond the laboratory processes that develop and modify the new giant molecules are the manufacturing techniques that bring them to their final form. These techniques, too, are evolving, to create forms and uses for the new materials which their original inventors could not even have dreamed of.

Fibridization, for example, is a process that has made it possible to turn nylon- and polyethylenelike materials into paper—and a type of paper with enormous advantages of wear and durability over all but the most expensive natural product. Normally, of course, paper is made from wood pulp—the natural polymer cellulose. Because normal manufacturing processes weaken the cellulose considerably, most paper deteriorates easily in moisture or high temperatures. But nylon as well as polyethylene paper resists both—even more effectively than expensive, rag-based natural papers. Today they are finding their major use as insulating materials in small electrical parts, like the coils and condensers in radio and television sets; their superior heat resistance permits small components to perform heavier service. Nylon paper, in particular, may prove to be the material of choice for important documents, guaranteeing them a longer life than those printed on cellulosic paper.

This new type of paper is produced by combining crimped filaments and small, fibrous particles of the synthetic, to create a material resembling a long, matted hank of hair. These fibrids, as the material is called, are then fed directly into a press, which flattens them into paper.

Specific polymers for specific jobs

Despite the awesomely complex technology that produces the raw materials of synthetic chemistry, the man-made giant molecules offer a kind of simplicity and versatility that natural materials do not. Before the advent of the synthetics, chemists and engineers had to design their products around the properties available in a relatively few materials, and they often had to use—and pay for—properties they did not really want. But with a broadening spectrum of polymers and copolymers, engineers can with increasing exactitude establish the precise properties they want in a material beforehand, and then turn to the polymer chemist to produce it. The changes wrought by that kind of freedom are not only creating better products, they are making these improved products more readily available at lower cost.

Plastics' New Bag of Tricks

Kapton, Merlon, Lexan, Mylar, RTV 615—these are members of the new generation of space-age substances: thinner, tougher, stronger and bouncier than any that came before. Lexan, for example, will not shatter or crack even when pummeled with a sledgehammer. RTV 615 *(opposite)* has the elasticity of rubber, but with two big differences: it is transparent, and it transmits almost no heat, even at temperatures four times hotter than the melting point of steel. Man-made materials with such properties are already being used in space vehicles, helping to protect Astronauts and spacecraft from the intense buffetings, pressure changes, and extremes of heat and cold to which they are subjected.

Adapted to more prosaic purposes, the new synthetics may produce some startling sights a few years hence—for example, houses made fireproof by protective layers of foamy silicone paint, and buildings without nails, held together by molecule-to-molecule welds formed by synthetic adhesives. There may come a day when people will live and work in undersea cities, protected by a filmy, translucent wall that permits oxygen—but not water—to pass through.

A COOL CAT
Perched atop a 4,000° F. flame, this kitten is protected from the heat by a slab of silicone called RTV 615, a transparent rubber that resists intense heat by an unusual process called ablation —a slow, layer-by-layer decomposition. Similar silicones have been used in the heat shields of space vehicles and have withstood reentry temperatures of 15,000° F. as long as three minutes.

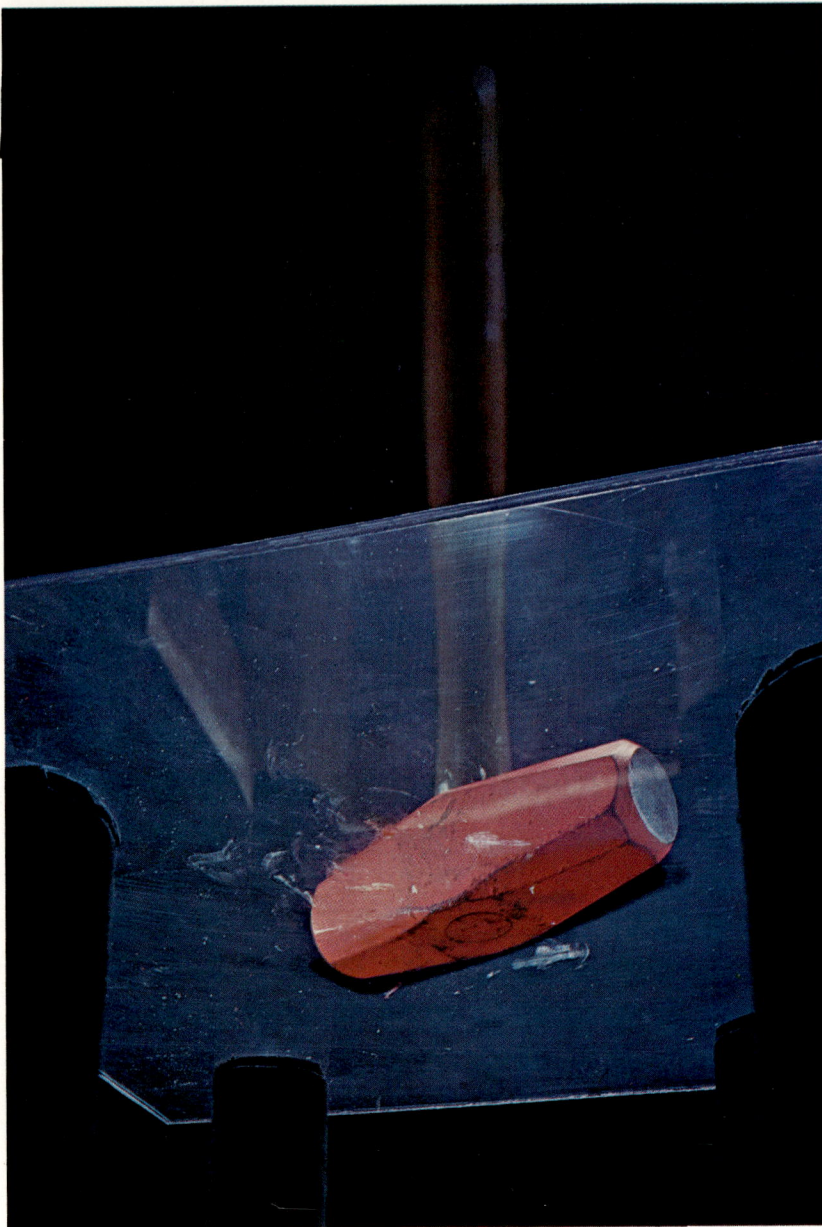

THE TOUGHEST PLASTIC

The 4½-pound hammer rests on a quarter-inch-thick slab of Lexan *(above)*, which resists the deadweight and gives harmlessly under the hammer's blows *(right)*. In another test, a thin sheet stopped a .38-caliber bullet fired from 12 feet away. Lexan and its twin Merlon are used in many products, ranging from the visors of Astronauts' helmets to vandal-proof windowpanes.

A Pair of Rugged Freaks

Possibly the most puzzling plastic in commercial use is a polycarbonate resin, sold as Lexan or Merlon, that is as clear as glass and nearly as tough as steel. Lexan was accidentally discovered in 1955 by a General Electric chemist, Daniel W. Fox, who was investigating polyester resins. Unable to remove his steel stirrer from a clear glob that had formed in one of his flasks, Fox finally broke the flask and tried to pound the substance loose —but no amount of banging seemed to affect the strange plastic. Fox was unaware, however, that two years earlier West German chemist Herman Schnell had made the same discovery, which was to become Merlon.

What gives these polycarbonates their unique properties is not yet clearly understood, but chemists believe it has something to do with the presence of ring-shaped groups in the long chain of the giant molecule.

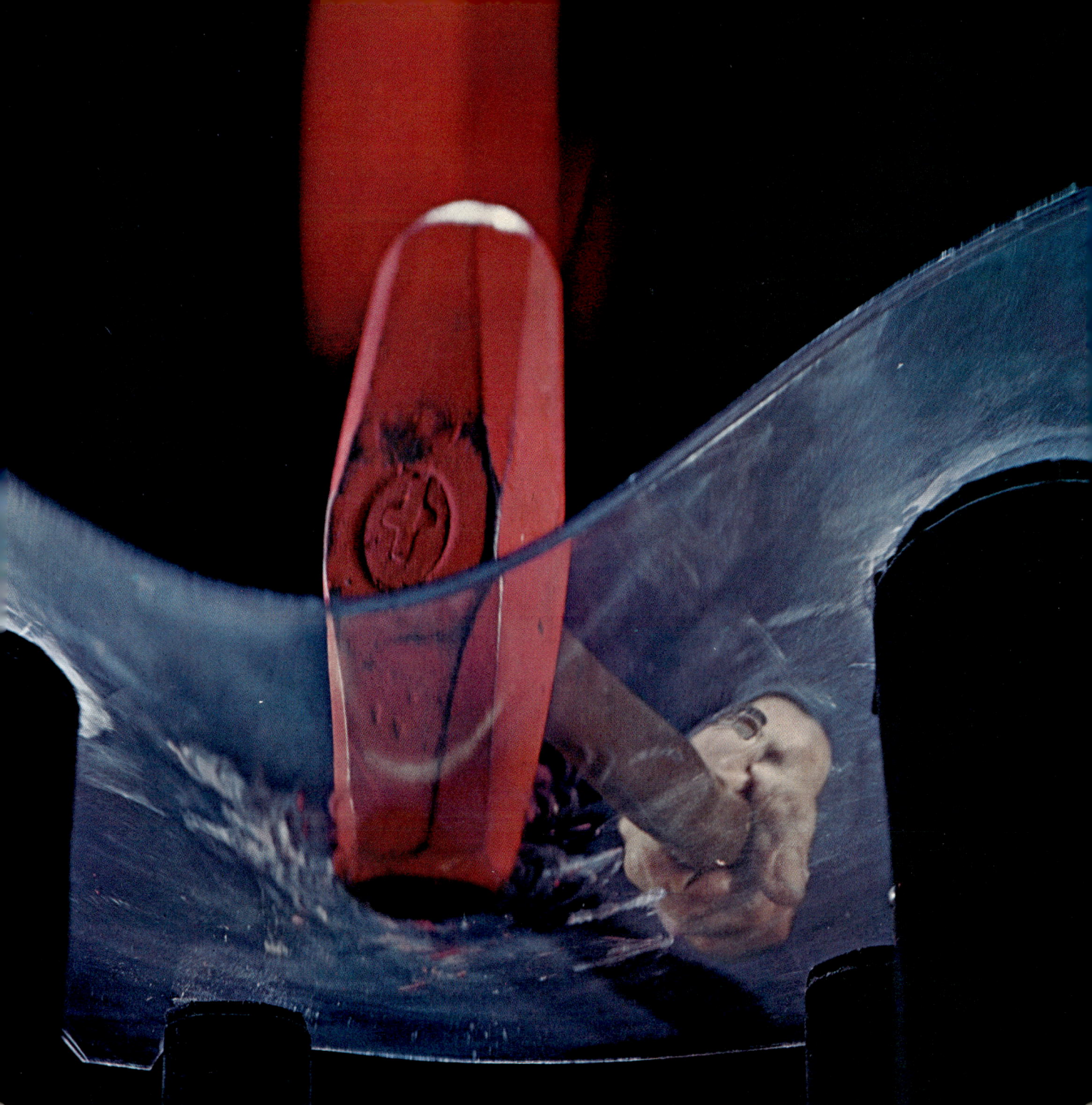

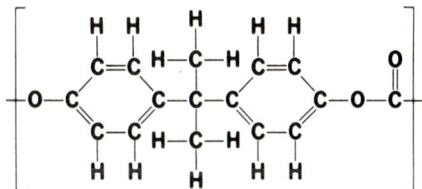

LEXAN

n = ~80-100

THE SHORTHAND OF CHEMISTRY

This diagram shows a Lexan monomer in the standardized shorthand used by chemists to reveal the content and structure of any substance. The figures below the diagram marked "n" indicate the approximate number of times such monomers must be repeated to form the giant molecule. Each letter represents one atom: C stands for carbon, H for hydrogen and O for oxygen—three elements found in most polymers. The lines between atoms tell how they are linked: a single line for a single bond, a double line for a double bond. Occasionally even a triple bond may appear. From the groupings formed by these atoms, and their positions on the chain, a chemist can deduce much about the properties of the resulting product. Thus, the hexagonal rings denote strength and rigidity. The presence of side chains—groups running off to either side—usually adds a degree of flexibility and toughness to the substance.

A Mighty Drop of Glue

It may be hard to imagine a world without nails, bolts, nuts, rivets and screws, but in years to come all of these may be replaced by plastic adhesives that are stronger, lighter and easier to use than conventional fasteners. A single drop of one such adhesive, Eastman 910, will bind almost any two materials into a virtually permanent union. While most adhesives require heat or evaporation to set properly, Eastman 910, applied as a liquid cyanoacrylic compound, forms a solid polymer on contact. The microscopic amount of moisture and oxygen present on virtually any surface is enough to trigger the on-the-spot polymerization. Tiny amounts of Eastman 910, applied on the point of a toothpick, are often used to bond hard-to-reach components of miniaturized electronic equipment, such as hearing aids and computers.

Eastman 910's strength results from the magnetlike attraction its carbon-nitrogen groups exert on the molecules of surfaces to be joined. In minutes the chemical weld created by one drop is strong enough to support a half-ton of weight (*opposite*).

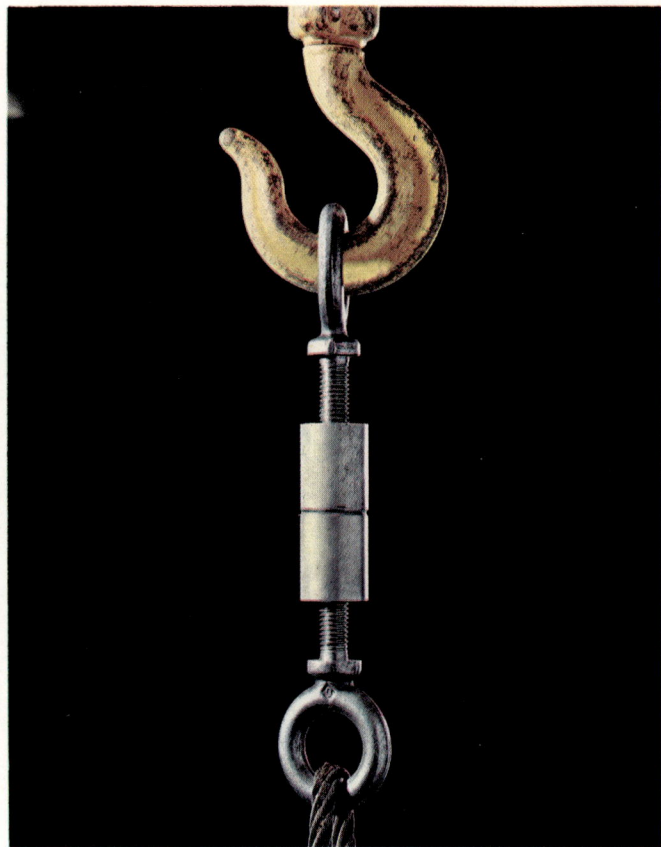

To test the sticking power of a single drop of Eastman 910, two steel cylinders are joined together to become part of a heavy-duty hoist.

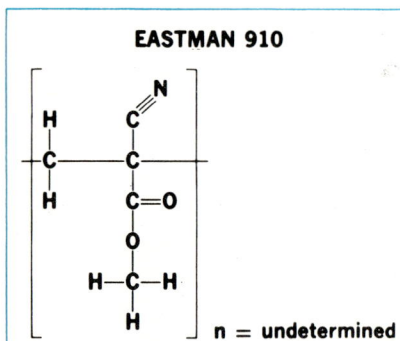

EASTMAN 910

n = undetermined

THE STICKING POINT

A triple-bonded carbon-nitrogen group (upper right in diagram) balanced with an ester—a carbon and two oxygens—produces stickiness.

A THIN PURPLE BOND

The Eastman 910 bond, even thinner than the two sheets of film it is joining together, can be clearly seen as a purple stripe in this photomicrograph of 1,000 magnifications. To the naked eye, however, the three layers appear as one.

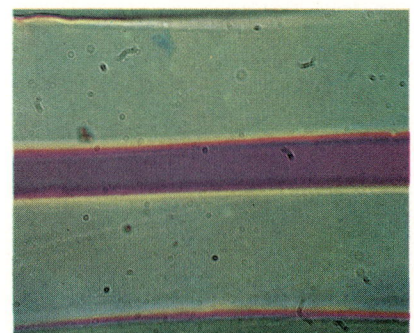

Tested under a rigger's watchful eye, the twin cylinders bonded with Eastman 910 support the full weight of a 1,200-pound safe.

BUTYL RUBBER

$$\left[\begin{array}{c} H \\ | \\ H{-}C{-}H \quad H \\ | \quad | \\ {-}C{-}{-}C{-} \\ | \quad | \\ H{-}C{-}H \quad H \\ | \\ H \end{array}\right]$$

n = ~100–20,000

NATURAL RUBBER

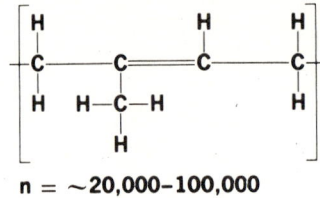

$$\left[\begin{array}{cccc} H & & H & H \\ | & & | & | \\ {-}C{-} & C{=} & C{-} & C{-} \\ | & | & & | \\ H & H{-}C{-}H & & H \\ & | & & \\ & H & & \end{array}\right]$$

n = ~20,000–100,000

CIS-POLYBUTADIENE

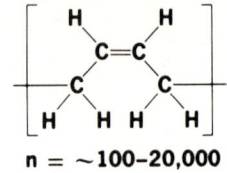

$$\left[\begin{array}{cc} H & H \\ | & | \\ C{=}C \\ | & | \\ C & C \\ / | & | \backslash \\ H \; H & H \; H \end{array}\right]$$

n = ~100–20,000

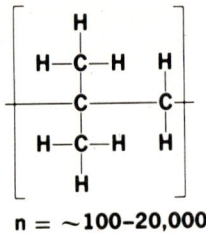

BOUNCELESS BUTYL
Butyl sacrifices resilience for impermeability—achieved by the two side chains *(top and bottom)* of a single carbon and three hydrogens.

BOUNCY NATURAL RUBBER
Natural rubber is more resilient than butyl, partly because only a single side chain *(bottom)* is fixed to the carbon backbone of its monomer.

BOUNCIER "CIS-POLY"
Cis-polybutadiene, much closer than butyl to the structure of natural rubber, is even more resilient; a key factor is its lack of side chains.

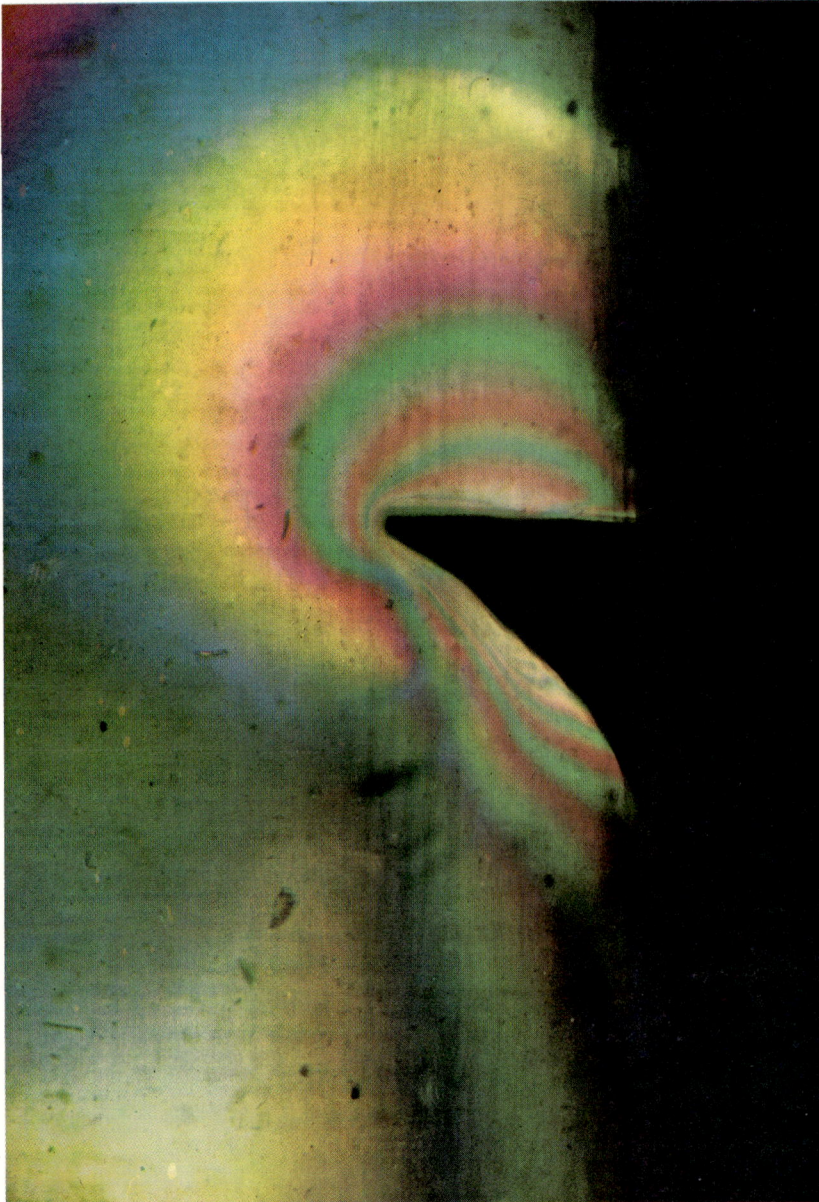

A World of Man-made Rubber

The American synthetic rubber industry was born early in World War II, when Japanese control of the natural-rubber resources of Southeast Asia created severe shortages. In the years since, the industry has grown into a billion-dollar giant. Today, about three fourths of all the new rubber in the United States is man-made, and the percentage is constantly increasing. From wartime processes has sprung a large, talented family of synthetic rubbers. It ranges from rugged butyl to bouncy cis-polybutadiene, a rubber specifically designed for tires.

Cis-polybutadiene, which was developed by the Phillips Petroleum Company, demonstrates how certain desired properties—in this case, abrasion resistance *(left)*, high resilience and low heat buildup—can be carefully built into a substance by polymer chemistry. A slight change in its structure produces trans-polybutadiene, a rubber hard enough to cover golf balls.

CLOSING RANKS
How cis-polybutadiene resists puncture is shown in this photomicrograph. The rainbow of rings around the pinprick indicates that the molecules have compressed into close-packed layers, making them more resistant to further penetration.

While a red butyl ball rolls dead, a blue natural-rubber ball keeps bouncing—and a yellow cis-polybutadiene ball bounces higher.

Resisting Flame and Frost

A plastic that ignores both ends of the thermometer, Kapton is a thin, yellow polyimide film that transmits heat *(right)*—but remains intact at temperatures as high as 752° F. and as low as −452° F., only a few degrees above absolute zero. Developed by Du Pont chemists, Kapton has unusual stability, which, along with thinness and electrical resistance, has made it an ideal electrical insulator for space vehicles. The lunar module that landed the Apollo astronauts on the moon contained more than 14 miles of Kapton-insulated wiring.

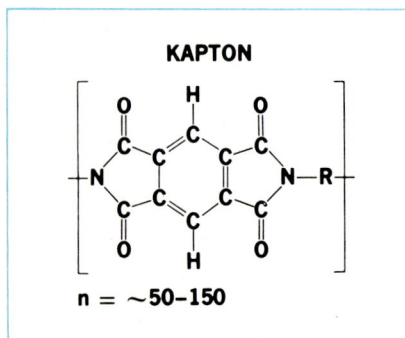

KAPTON

$$n = \sim 50\text{-}150$$

A CHAIN OF RINGS
Kapton, whose monomer consists principally of three interlocking rings, owes its stiffness and thermal resistance to its molecular structure.

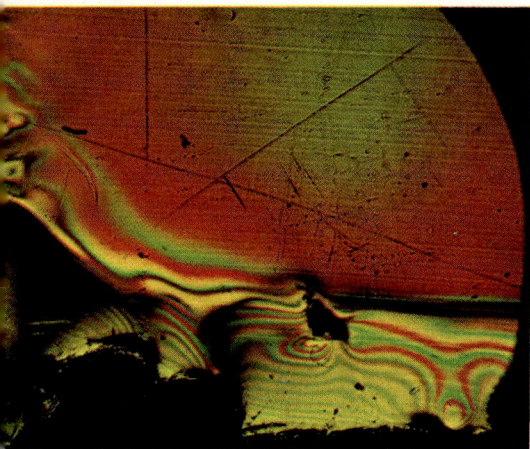

A FIERY TEST
A flame boils water but a $1/1{,}000$-of-an-inch-thick Kapton film between them remains intact. At 1,472° F., Kapton will char but will not burn. The yellow tint, also evident in the photomicrograph above, stems from the filtering out of blue light, a phenomenon not fully understood.

Surrounded by silicone, a canary peers unconcernedly at passing fish. The bird is sustained by oxygen passing through the membrane.

A Filmy Wall That "Breathes"

Fish can breathe by drawing in oxygen through their gills, membranes that filter dissolved gases from water. Now, with the development of an ultrathin silicone rubber membrane that is, in effect, a synthetic gill, man may have an ingenious device for underwater endeavors. The membrane is formed of two laminated layers of silicone film that, combined, are only one thousandth of an inch thick (*below*) and is completely watertight, but permits the passage of gases through its molecular structure. Thus, a submarine with rubber silicone windows on its inner skin would no longer have to carry bulky air tanks but could draw oxygen from circulating sea water and expel carbon dioxide by the same route.

Silicone film—made of polymer molecules that are derived from common minerals like sand—has long been known for its gas-permeable qualities. But only recently have scientists been able to produce silicone film thin enough to permit oxygen to pass through in life-sustaining quantities. The membrane is now being used in artificial lungs during open-heart surgery, performing the lungs' function of oxygenating blood and eliminating carbon dioxide. Because it transmits oxygen faster than some other gases, it can also enrich the air breathed by hospital patients.

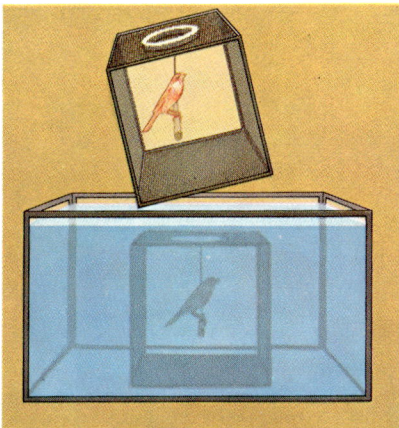

ONLY A BIRD IN A SILICONE CAGE
Putting a canary into a fish tank requires a one-cubic-foot, wire-mesh cage lined with silicone film. Completely watertight, the cage transmits enough oxygen to sustain the bird indefinitely, provided the water in the fish tank is kept fresh. For a human, the cage would need 260 square feet of film—equivalent to an eight-foot cube.

THE SECRET OF SILICONE
Silicon and oxygen form a highly flexible backbone in the silicone monomer, permitting gases to diffuse through at varying rates.

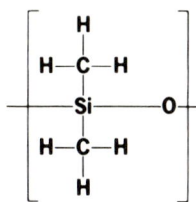

SILICONE RUBBER

$$n = {\sim}10{,}000\text{--}40{,}000$$

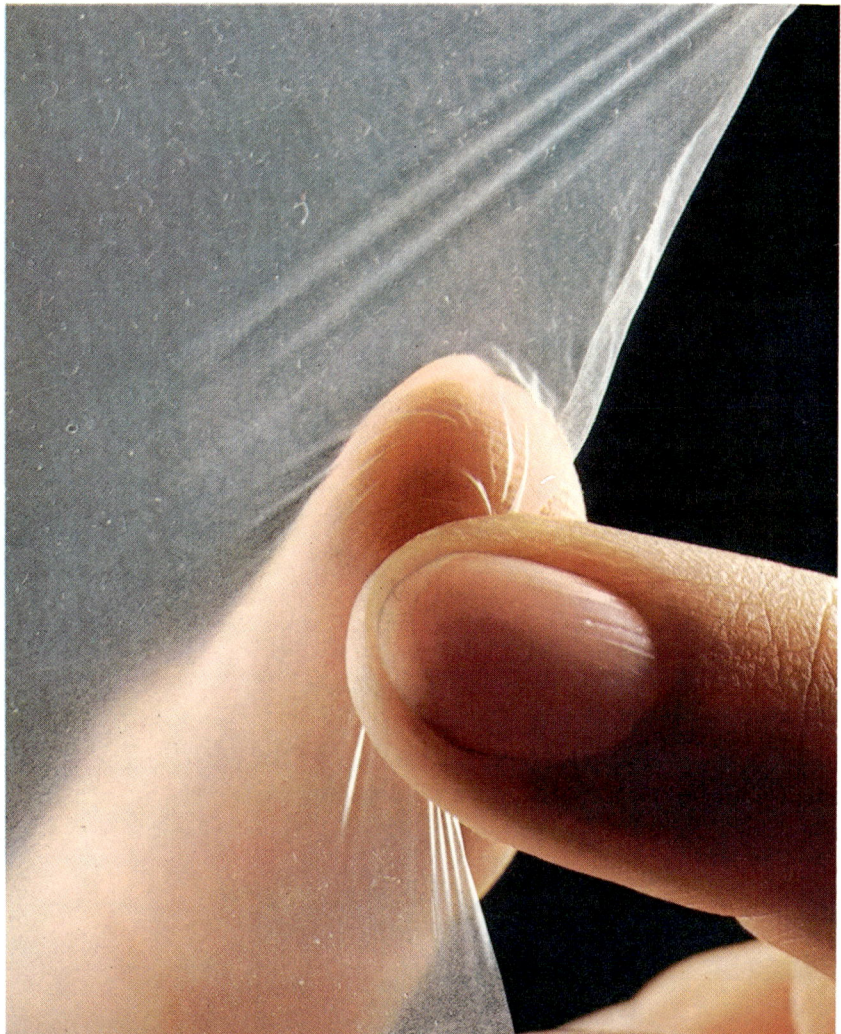

THINNER THAN SKIN
Silicone, to form a selectively permeable membrane, is made 1/2,000 of an inch thick, and two sheets are laminated together to ensure the holeless film shown here. Used now to sustain human organs in storage for transplanting, this type of silicone membrane might, someday, become an integral part of lung transplants, aiding the implant in carrying out its gas exchanges.

165

A Film of Many Talents

Of all the plastics now marketed, one of the most familiar and most versatile is Mylar. It can be magnetically coated as tape for sound and TV recordings, processed to make colorful coatings for paper and leather; or shaped into highly durable packages for frozen foods. A polyester sheet more flexible and far tougher than the cellophane it resembles, Mylar can be produced as a film 13 times thinner than a human hair. But even then, it retains the toughness shown by the sample at right: only 1/1,000 of an inch thick, it was impossible to pierce with a hard-thrown baseball.

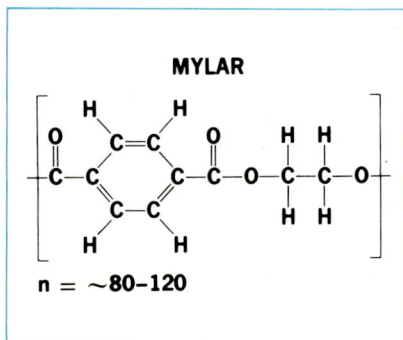

A SOURCE OF STRENGTH
The double-bonded carbon and oxygen atoms fuse Mylar's molecules into close-knit layers; 4,000 giant molecules in 1/1,000 of an inch.

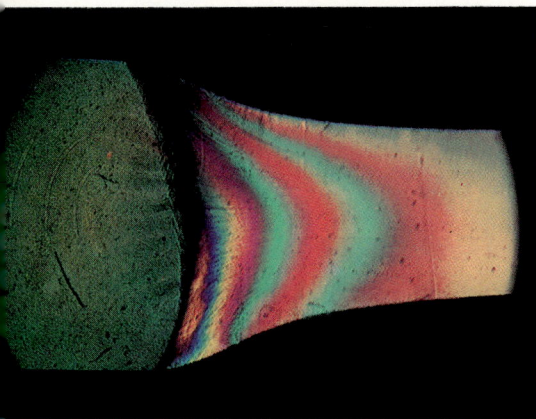

STOPPING A FAST BALL
A sheet of Mylar absorbs the impact of a baseball thrown at 80 mph *(right)*. The rainbow-colored linear formations in the photomicrograph *(above)* show the film under a similar stress. They indicate the patterned alignment of the molecules as they absorb the shock.

166

8

Wrapping
a City, Replacing
a Heart

A powerful scientific tool, this particle accelerator produces radioactive atoms, useful in discovering new details of polymer structure.

FOR ALL ITS GLITTERING ACHIEVEMENTS to date, the science of giant molecules is still young. One of its founding fathers, in fact, fondly likens it to an ungainly adolescent, ever exploring, improving its skills, seeking where next to test its strength. Even a glance at its explosive growth is enough to suggest that the young stalwart's most spectacular accomplishments are yet to come.

In the single generation or so since World War II, giant molecules have become the object of one of the most intense investigations in the history of science. Just before the War, there were fewer than 1,000 men in the United States who could call themselves polymer chemists. Today some 30,000 of these highly specialized scientists probe and manipulate new molecules in hundreds of industrial and university laboratories. It has been estimated that a new compound (but not always a valuable one) is now synthesized on an average of every five minutes. One result of the chemical industry's high volume and competitive technology is that most synthetics are getting cheaper, and many of them are improving in quality and versatility as well. New blends, hybrids and composites are evolved almost daily, combining the strong points of their various compounds.

Significantly, however, not all of this activity is aimed solely at weaving ever more miraculous, wrinkleproof fabrics or at building better plastic mousetraps. Constantly the discoveries of the polymer chemist draw him closer to the paths of other disciplines: space science, architecture, agronomy, oceanography, computer design, medicine, biology, civil engineering. It is from these collaborations that polymer chemistry promises to bring forth its most radical and valuable contributions to the shaping of man's environment and, in all probability, of man's body and brain as well.

By projecting principles and techniques at least partially understood today, it is possible to visualize some astonishing possibilities for polymer technology over the next 10, 30 or 50 years: inexpensive housing, including foamed varieties that virtually construct themselves; plastic membranes that filter seawater or sewage to supply water-starved cities and deserts; superefficient, all-plastic engines, fuel cells, solar-energy collectors, automobiles, boats, airplanes, rapid-transit trains; information-handling systems that pack whole libraries in filing cabinets and reduce room-sized computers to hand-portable machines; "detergents" that wash the infected blood clean of dangerous deposits before they can harm the body. Some chemists dream not only of replacing any diseased body part with a better synthetic one, but also of controlling disease, genetics and the synthesis of life itself.

If science follows its often erratic patterns, the realization of these and other dreams will probably come in unexpected ways. Certainly it will depend on a number of sociological, economic and other factors that are out of the polymer chemist's hands. But it will also depend on the chemist's own progress toward some elusive technical goals. Among these goals are plastics that are much stronger, more resistant to temperature extremes, quicker to polymerize and easier to produce in large quantities.

Strength is a matter of immediate concern. Because chains of molecules start forming at different times under the influence of catalysts, most polymerization processes today produce a wide range of molecular weights within a given substance, including many short chains that make the mixture structurally weak, and many long ones that make it hard to process because of high viscosity. To solve these problems today, the chemist can only work backward, extracting these wrong-sized molecules after the mixture has been made. But if he could start all the chains building at once, keep the conditions of growth constant, and stop them all after a short, controlled period, he could produce a polymer consisting solely of ideal strands of medium size and relatively uniform molecular weight. He could then extrude far stronger plastic pipe or blow far tougher film. He could also do the job much faster in higher-pressure equipment, approaching the production speeds now achieved with metal. By means of such techniques it might even be possible, for example, to mount highly efficient extruding equipment on a truck or barge and lay cheap, continuous, seamless pipe under ground or water. The implications for industrial pipelines or desert irrigation might be broad indeed.

To make plastics strong enough to do new jobs, the polymer chemist presently has at his disposal various chemical and irradiation methods that cross-link overlapping molecules into a tighter knit. But some plastics still tend to elongate under tension like taffy, or crack under bending or impact, or turn soft under heat. Reinforcing them with strong fibers of steel, glass or other materials—much as concrete is reinforced with metal rods—may remove most of these flaws. This reinforcement provides the strength and resiliency characteristics of the glass fiber—actually a combination of plastic and glass fibers—used for fishing rods, boat hulls and the plastic body armor developed experimentally for foot soldiers. By manufacturing even stronger reinforcing fibers out of pure materials without structural defects (e.g., small crystalline "whiskers" of silicon carbide or boron nitride fibers), it has been possible to make certain plastics strong enough to outperform metals for such vital uses as spaceship hulls and the wings of supersonic airplanes.

Making a ship with a spider's thread

But the polymer chemist regards even these achievements as stopgaps. If he could somehow increase the length of the molecules themselves so that they reached throughout the entire object—fiber, film, rod or pipe—he might achieve the legendary superplastic that would never break. In some of his fondest giant-molecular dreams, the polymer chemist sees himself standing at the ocean's edge, spinning his heavenly yarn into ship hawsers the diameter of spider's thread and into bridge cables that send a single, continuous molecule soaring toward misty shores a dozen miles away.

Somewhat closer to realization is another dream: truly high-temperature-resistant polymers that cost one dollar a pound or less instead of the present price of over $20 a pound. Until a few years ago, no polymer could withstand heat much higher than 650° F.; today government contracts require special plastics that protect metals from temperatures as high as 4,000° F. for short periods. Used as coating on the inside of rocket-engine combustion chambers and thrust nozzles, and on the re-entry end of space capsules, these plastics break away or evaporate slowly on the surface exposed to heat, cooling the rest of the material and protecting the shielded metal. Such plastics are made up of chain molecules that have a benzene-ring-pattern structure similar to that of pure graphite. However, the molecules are arranged in linear bundles to give them the strength and pliability of a fiber. If the raw materials for such plastics could be made available cheaply and in quantity, and the processes could be perfected to shape them into foams, tubes, plates and rods, the plastics would have much broader uses. They could be employed as building materials that would never burn or melt; as surface protection for metal parts—or as the parts themselves—in lightweight, corrosion-proof gasoline engines or turbines (which would then need no lubrication if the plastic was also made frictionless with fluorocarbons); as insulation that would drastically reduce the size of electric motors of all kinds by allowing much more current to be pushed through a given coil without buildup of destructive heat.

The goal: swift synthesis

Closest to the heart of many chemists is the dream of speeding up the polymerization processes. Compared to growth processes in plants and animals, which are measured in months and years, synthetic polymerization is already fairly rapid—a few hours for nylon, a few minutes for polyethylenes, a few seconds for fast-setting adhesives. But if polymerization could be made to occur in only fractions of a second, important new developments would be possible. Production would be simplified: filaments could be spun, parts extruded and films blown directly from monomers, with no costly shipping, mixing and batching stages in between. Monomeric inks, polymerizing instantaneously and drying on contact with paper, could make four-color printing faster, cheaper and crisper-edged. Photography itself, imitating photocopying, might be based on polymerization. Such a system, requiring only a single film, might replace even multilayer wet processes such as Polaroid, which produces in-the-camera developing in 10 seconds or less.

Closely related to photography are the intriguing potentials of photochromic (and thermochromic) materials, substances that change from colorless to colored on exposure to light (or heat) and back again when the stimulus is removed or another stimulus employed. Photochromism has been observed in several hundred organic and inorganic compounds, but the mechanisms of relatively few have been fully worked out. They involve changes in atoms or molecules that rearrange themselves structurally when certain wavelengths of light hit them, and later switch back to their original shape. Photochromic plastics have already been incorporated in goggles; the plastic becomes dark faster than a human can blink, protecting military pilots' eyes from blinding nuclear blasts. Many photochromic materials are unstable and wear out eventually, but improved varieties are being marketed in such novelties as "suntanning" dolls, and as "automatic" sunglasses which adjust their tint from light to dark as the wearer steps from indoors into the sun, and back to almost colorless when the sun goes under a cloud. Other new applications include automobile windshields that darken not only against dazzling sun but also against oncoming headlight glare. Glass companies, which have already perfected heat-absorbing and "one-way" tinted glass for buildings, are trying to find materials cheap and long-lived enough to put into ordinary windows and thus eliminate the need for curtains and blinds.

The Bible on a two-by-two card

To some technicians, however, the most exciting role for photochromics lies in serving increasingly complex information-handling needs. Because they have no "grain" like conventional photographic chemicals, certain photochromic dyes have made possible supermicrofilming that reduces images to 1/40,000 of their original size. Incorporated in an organic polymer on a piece of glass, these colorless dyes darken almost instantly when an optically reduced image is projected on them in ultraviolet light. By use of this method, a 1,245-page Bible has been photomicroimaged, page by page, across a single transparency less than two inches square, and a 2,580-page U.S. Navy parts catalogue is stored on one reproducible three-by-five-inch card. Viewing equipment to project this incredibly miniaturized information back to readable size has itself been made small enough so that pilots or Astronauts, for example, can carry 50,000 pages of navigational charts or technical procedures in a hand-held screening machine no larger than a portable radio. The implications for corporate records, libraries and teaching systems are astounding enough, but some scientists are even more enthusiastic about the potential for data-processing. By optically storing millions of bits of information per square inch in small stacks of photochromic film, huge and costly computers might someday be reduced to handy and widely usable tools no bigger than electric typewriters or adding machines.

Whether or not a man will soon carry around his own pocket-sized libraries and computers, there is little doubt that polymers will continue to play an increasing role in almost everything made by man. They are being made light, durable, strong; moldable into virtually any shape; inert to corrosion of most kinds; resistant to temperature extremes; chemically "slippery" enough to slough off dirt or oil (and to act as their own lubrication in moving parts), or sticky enough to hold together a wide range of structural parts with high-strength, continuous bonds.

The growth of high-strength adhesives suggests that someday even the screw, the rivet and the nail will be looked back on as quaint devices of the past. Houses, cars and planes, for example, may someday be completely glued together as furniture is today. During World War II, the Allies' lightest and fastest bomber, the British-Canadian Mosquito, was constructed almost entirely of those old-fashioned polymers, doped canvas, plywood and glue; today's foremost all-weather fighter-bomber, the American scissor-wing F-111, uses newer polymers to bind together its metal wings, tail, fuselage and skin against the stresses of speed two-and-a-half times that of sound. Filament-wound or fiber-reinforced plastics are replacing aluminum and even titanium in the bodies of some airplanes and rockets where strength-weight-temperature characteristics are critical. Plastics are making progress on the ground as well. Bit by bit, the amount of plastic in automobiles being produced increases. Also, some designers speculate that if rapid mass transit is ever to be achieved in crowded interurban areas, lightweight plastics must provide the answer, if only to reduce the power needed to move trains at speeds of 200 miles per hour or more.

A city encased in plastic

Meanwhile, plastics are steadily moving into the large, slow, everyday business of building. Two of the more celebrated big buildings of the space age are, in fact, built largely of plastics. The huge, boxlike rocket-assembly building at Kennedy Space Center is brightened by 418-foot-high translucent walls of reinforced polyester resin panels coated with polyvinyl fluoride, chosen for glare-free working conditions, durability, and resistance to breakage by hurricanes and launch-pad shock waves. The 750-foot-diameter Houston Astrodome, the first fully enclosed sports stadium in the U.S., is roofed by a double, insulating sandwich of acrylic panels. Some architects dream of ways to remove the supporting steel now needed in such large, rigid domes; they picture athletic arenas, shopping centers and other superbuildings roofed by vast, air-supported plastic balloons—as in fact some smaller winterized tennis courts, swimming pools and traveling exhibits now are. From there, a few forward-thinkers have jumped to concepts of the plastic-enveloped city, under whose air-conditioned firmament there would be neither rain nor snow, heat nor cold, only flowers, fountains and civic joy. Meanwhile, plastics continue to meet more pressing civic demands. At least one big-city school, troubled by constant vandalism, has replaced all its window glass with clear acrylic, and the rocks now bounce harmlessly off, to the frustration of would-be vandals.

Of all the possibilities of plastics in building, however, one of the most inherently fascinating to designers is that of foams. Flexible foam rubber mattresses, of course, have been around for quite a while, as has rigid foam insulation inside building walls, refrigerators and even nuclear submarines and high-altitude aircraft.

Regardless of their application, all plastic foams are essentially cellular structures composed of millions of gas bubbles separated from each

A POLYMERIC BIBLE, the miniature set of Scriptures shown above—all 1,245 pages of which are contained in a space less than 1 3/4 inches square—was photographically reduced on plastic film coated with a special, light-sensitive dye. Magnified 48,400 times, the text appears full-sized *(below)* with almost no loss of definition. According to one estimate, all the books in the Library of Congress, reproduced by this method, could be stored in six ordinary filing cabinets.

other by thin walls of plastic film. They can be made with several different polymers, using different methods to achieve different results. In one of the most common, urethane foam, excess isocyanate groups in the urethane polymer react with water or acid to make carbon dioxide gas, which "blows" the foam up to 30 times the volume of the original material while the molecules cross-link and become rigid. This forms a strong, lightweight substance whose trapped bubbles make it an excellent insulating material. Foams can be produced as planks, slabs and other carefully controlled shapes in the factory, or sprayed or blown into building forms in the field. Moreover, if the containing form is cooled, foams tend to make their own tough, bubbleless surface skin, and often need little or no further finishing. Prefinished planks of foam have been laid in tight, continuous spirals to create self-supporting domes up to 80 feet in diameter for small theaters, convention halls and other buildings.

The lightness, versatility and portability of foam structures have made them obvious candidates for such special uses as camp shelters, vacation houses, military barracks, field hospitals, communications centers and maintenance shops, as well as for temporary or disaster housing of various kinds. Aid agencies are considering the use of inexpensive foams and other building plastics to alleviate serious housing shortages in developing countries. In all such applications, designers are intrigued with the possibility of making foam shelters not only easily erected and portable but otherwise self-contained: rather than relying on expensive (or nonexistent) power, water, sewage and telephone networks, the plastic house of the future might well use plastic fuel cells, chemical toilets and wireless radiophones.

A house on the moon

These concepts of a light, self-contained shelter have figured prominently in the thinking of space technicians who have been concerned with the problems of living on the moon. Self-expanding and rigidifying polymers have been used to form the dish-shaped solar-energy collectors that operate in space to power weather and communications satellites. The Air Force and industrial contractors have tried similar systems for producing igloolike shelters that are chemically or electrically triggered to inflate and rigidify from a relatively small package. Plastics are also being tested as membranes to separate the hydrogen and oxygen compartments of fuel cells that power space capsules, and in systems that purify for reuse the limited supplies of air and water Astronauts can take along with them on their space travels.

Plastic membranes are also under study for broader applications on earth. Chemists may well improve on present charcoal filters and resinous water-softeners with new polymerized adsorbents capable of removing the noxious gases from smokestacks and auto exhausts before they form dangerous urban smogs, and the chemicals from industrial wastes before they pollute the general water supply.

Of even broader significance for water-short urban and rural areas are membrane processes which purify sewage water or desalt ocean and brackish inland water at a much lower cost than distillation. Much research is currently going into two processes involving such membranes: electrodialysis, in which electricity is used to pull the salts from the water that passes through the membrane, and reverse osmosis, or ultrafiltration, in which the membrane itself filters out the salt. Salt might be removed from seawater by means of such methods for less than 25 cents per 1,000 gallons, a fraction of the cost with conventional desalting systems. Furthermore, the valuable concentrations of sodium, potassium, magnesium, bromine, chlorine, copper, silver and gold contained in seawater might well be profitable by-products.

Growing trees in bags

Just what down-to-earth solutions polymer chemistry may be able to offer to mounting world problems of food and water is not easy to predict. In some drought-prone areas, thin layers of paraffins and polyethylene have been used successfully to coat the surface of ponds and reservoirs, thus reducing evaporation of scarce water supplies. In other experiments, scientists have enveloped fruit trees and plants, roots and all, in transparent polyethylene bags, forcing them to recycle a limited amount of precious water in a closed system without appreciable loss. Some scientists visualize highly controlled food production in huge plastic-roofed greenhouses; others picture big algae farms or tanks of microorganisms producing high concentrations of needed proteins for use as food additives when dried.

In the realm of pure synthesis, chemists have made certain simple, edible polymers in the laboratory. But the cost cannot begin to compete with the efficient, natural protein-producing process of animals and plants. Polymer chemistry's most notable contributions to food technology so far have been in packaging and chemical preservation for shipping, and in the pure synthesis of supplements to make food more nutritious. Indeed, the construction of new factories for production of such amino acids as lysine suggests that the deficient plant proteins of the world's staple grains—corn, rice, wheat—could be made as rich as steak.

Will chemistry ever be able to synthesize complex substances such as steak itself? The answers are not easy to come by. In the first place, proteins are among the largest of the giant molecules, with molecular weights ranging into the thousands and even millions. For example, the molecule of the human blood protein, called serum albumin, has been found to consist of 526 amino units of 18 different types with a total molecular weight of about 69,000. But the properties of the molecule depend on its structure—how these amino acids are arranged. The number of possible arrangements in which only 18 different amino acids can be placed is roughly six quadrillion; the number of possible arrangements of 526 different units is inconceivably large. Protein chemistry is still in its infancy, and chemists have not yet been able to build up long chains with amino acid sequences like those found in animal proteins. But they

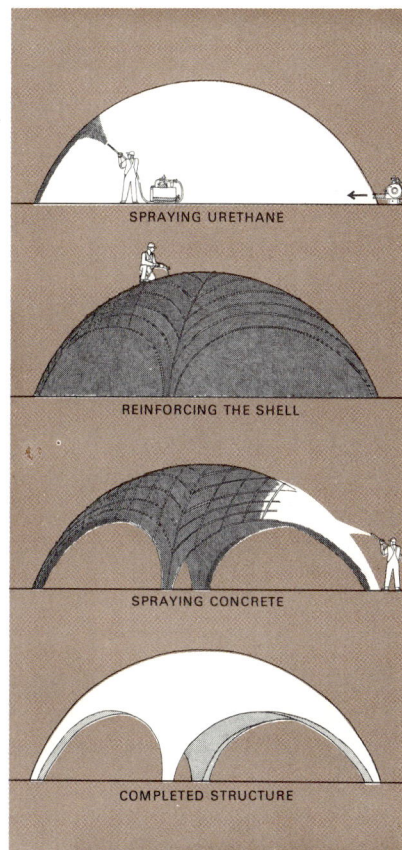

SPRAYING URETHANE

REINFORCING THE SHELL

SPRAYING CONCRETE

COMPLETED STRUCTURE

A PLASTIC BUILDING SHELL, this dome and others like it enable builders to construct buildings without framework or supporting pillars. In the proven technique shown here, a two-inch coat of urethane foam is sprayed onto a form of plastic film held taut by compressed air. After drying, the light urethane shell is covered with metal reinforcing bars and openings are cut out. The outer surface is then sprayed with three inches of concrete. The result is a blend of modern design and durability. Using urethane foam gives the builder an added dividend: it is an excellent insulating material.

have polymerized single amino acids into chains from which textile manufacturers have spun natural-protein fibers almost identical to silk (and, so far, just as expensive). Chemists would like to know why the molecular structure of natural fibers gives them their special properties—why it is, for example, that although cotton feels warm and absorbs moisture, those properties are not usually found in synthetic fabrics.

Human skin from a test tube

From duplicating natural fibers to creating living tissue is an enormous jump; skin and flesh are highly complex matter that perform an almost infinite number of tasks. But some chemists think that this synthesis too may come in some form, perhaps in 20 years or less. The starting point for any synthetic skin, as for other proteins, would doubtless be some polyamino acid; the end product might revolutionize plastic surgery, speed up cut- and burn-healing and virtually eliminate the painful cutting and grafting of real skin. One problem, of course, is to make synthetic skin nontoxic and incapable of arousing reactions that would cause the body to reject it.

In this area, much experience has been gained in designing plastic substitutes for other parts of the body. Silicone rubbers and other chemically inert, pliable polymers are now used successfully for everything from artificial eye corneas to heart valves, and even a rudimentary version of a lung. The versatile silicone rubber Silastic, in fact, can be made into a film envelope filled with a silicone gel to imitate body fat in a synthetic breast; since it has the texture and resilience of soft tissue, it can also be made to resemble muscle in an artificial heart, or cartilage in ear and nose replacements.

Polymers are also playing increasingly important roles in body metabolism. Cellophane membranes are used outside the body in artificial kidney machines to diffuse waste products from the blood and deposit them in a reservoir for removal; improved membranes promise to duplicate more closely the sophisticated ultrafiltration process of the kidney itself. Medical engineers hope eventually to make an artificial kidney small enough to fit inside the body, much as they have implanted plastic-encased electrical "pacemakers" in thousands of patients to regulate or restore the rhythm of an ailing heart. Scientists foresee the complete replacement of a diseased heart with a nonclotting, long-wearing synthetic one, and the possibility of hooking up heart patients periodically to blood-laundering machines, which would use polymeric absorbents to remove specific materials such as fats or calcium compounds from the bloodstream before they can form harmful atherosclerotic deposits. Finally, the affinity certain polymers display for other molecules may one day be put to work right in the bloodstream, either in the form of "detergents" which could combine with harmful matter and precipitate it out of the blood into the urine, or as selective agents capable of carrying specific drugs to specific parts of the body without depositing them in others, thus eliminating the dangers of poisonous side effects.

Perhaps most dramatic of all the possibilities are those chemistry now holds out for the control of life itself. The most exciting new field of chemistry is that of molecular biology, whose very name suggests a tribute to the prime and lasting concept of the molecule. Slowly scientists are analyzing the structures and roles of the molecules that make life possible —proteins, enzymes, hormones, nucleic acids, antibodies—and are trying to unravel the mystery of how nature puts them together and how they act. Already some of these vital giant molecules can be synthesized. The first man-made enzyme, ribonuclease, was produced in 1969, and if this feat can be repeated for other enzymes, medicine may be able to treat many deficiency diseases and perhaps even come to understand what it is that makes cancer cells behave differently from normal cells.

When chemists have uncovered the secrets of the DNA (deoxyribonucleic acid) and RNA (ribonucleic acid) molecules, which control the genetic coding and manufacture of the body's cells, the implications may be fantastic indeed. If man could control genetic coding, he could control body and brain size, skin color, stamina, intelligence, mental health. Some experiments with animals indicate that learning capacity may be transferred from one living body to another by taking RNA, the so-called "memory molecule," from one brain and injecting it into another through the bloodstream. Body and skin to order, education by injection, even suspension of the aging process in muscle and skin—serious scientists are actually discussing ideas as astounding as these.

The man-made man

In 1967, scientists at the Stanford Medical Center in Palo Alto actually synthesized a virus DNA, seemingly indistinguishable from the DNA of an active virus. This great achievement brought closer to reality the radical proposal of Charles C. Price of the University of Pennsylvania. Retiring as president of the American Chemical Society in 1965, he said, "We may be no further today from at least partial syntheses of living systems than we were in the 1920s from the release of nuclear energy, or in the 1940s from a man-in-space. The political, social, biological and economic consequences of such a break-through would dwarf those of either atomic energy or the space program. Success could lead to modified plants and algae for synthesis of foods, fibres and antibodies, to improved growth or properties of plants and animals or even improved characteristics for man himself."

Scientists do not know where or when they may stumble upon the means of achieving these miracles, or what the full implications may be. But researchers have rarely known; the history of science is replete with fortunate accidents and inspired, but often unexpected, flashes of insight that have brought about momentous changes in the world. The search for understanding of the physical laws that govern the universe is endless. At the moment, the polymer chemist is certain of only one fact: he has crossed the border of another great scientific frontier whose dimensions are incalculable.

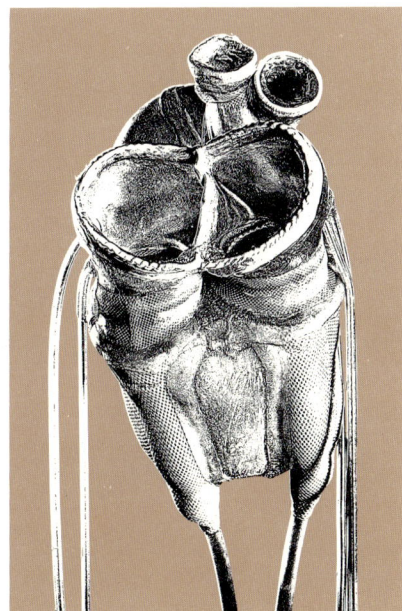

A PLASTIC HEART, made of silicone rubber, has kept a calf alive for 96 hours. Researchers are also experimenting with plastic lungs and muscles, in the hope that these devices can be used to replace diseased body parts. Less complex plastic replacements, like synthetic heart valves and artery sections, have remained in patients for years with excellent success.

The Gas That
Turns into Leather

Once research has created a new synthetic, it is up to industry to transfer the process from the test tube to the production line. Sometimes the transition requires years of construction and experimentation, and the expenditure of millions of dollars. The result is often a system of complex plants in which the manufacture of the product is almost completely unseen: the basic raw materials go in at one end in one form and come out as something altogether different.

An example is the synthetic leather called Naugahyde. In the laboratory small quantities of the polymer could probably be made at relatively small expense. Producing it in quantity, and making it into a useful material, is another story. Naugahyde is the result of a process that involves three plants hundreds of miles apart, as well as a number of dramatic but invisible chemical actions. First, natural gas is converted to highly volatile acetylene, then the acetylene goes into making a liquid monomer, which in turn becomes a granular solid, polyvinyl chloride. Finally, huge rollers coat the vinyl onto a fabric backing—and out rolls Naugahyde, a material more durable and adaptable than leather.

THE STARTING POINT

A worker turns up the steam pressure of the boilers at the Monochem plant in Geismar, Louisiana, beginning a process that will ultimately convert natural gas into Naugahyde, a man-made leather. The steam is used to drive air through compressors and into a cooling unit, where liquid oxygen (used to strengthen the combustion of natural gas) is separated from the atmosphere.

Sorting Out
the Gases

The process of making Naugahyde starts in Geismar, Louisiana, a tiny town on the Mississippi River near Baton Rouge. Here Uniroyal and the Borden Company jointly maintain a highly automated plant, employing only 130 people, to produce the vinyl chloride monomer—a crucial early step in the production of Naugahyde.

From beginning to end, the Mono-chem workers never see what they are making. The initial step is the breaking down of natural gas—piped in from nearby gas fields—into molecules of acetylene, carbon monoxide and hydrogen. This is done by heating natural gas to 2,700° F., a temperature achieved by burning it with oxygen. The separated gases, quickly trapped and purified (right), are all put to commercial use. But only the acetylene goes into the monomer.

A LIQUID COLDER THAN ICE
The boxlike towers above produce liquid oxygen by chilling air to −290° F. At this temperature, the oxygen liquefies and can be separated. A technician at the oxygen plant (right), his face and hands protected from the supercold liquid, draws a sample into a Thermos before the oxygen is vaporized and propelled under pressure through a pipe network to the acetylene plant.

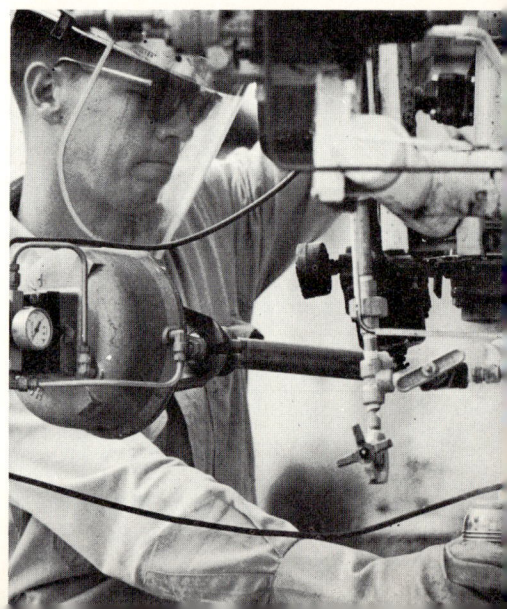

A MOMENTARY INFERNO
In the acetylene plant, burning gas wastes cast brilliant flares high into the sky. The natural gas and oxygen are preheated to 1,100° F. before being mixed and ignited in furnacelike reactors. The resulting hot flame is immediately doused with cold water to keep the highly flammable acetylene, formed by a rearrangement of the natural gas molecules, from burning off.

AN AUTOMATED FLOW
Checking the rate of flow of vinyl chloride into a purification tower, two technicians examine recorders linked to various valves. All of the Monochem processes are controlled at four such centers, from which every phase of the plant's operation can be charted and directed.

Making a New Molecule

Acetylene, a very explosive gas, must be combined with hydrogen chloride to produce the vinyl chloride monomer. The process begins in sturdy gas mixers like the one at left, into which the acetylene is forced after passing through a low-pressure pipe system. With acetylene, the pressure may never exceed 25 pounds per square inch. Greater compression will cause the gas to explode.

From the gas mixer, the acetylene-and-hydrogen-chloride mixture—still highly flammable—is piped into one of the huge reactors below. There a catalyst, mercuric chloride, is introduced that causes the atoms of the two gases to react and combine to form a new molecule, with a fairly simple structure consisting of three atoms of hydrogen, two of carbon and one of chlorine, and a double bond connecting the carbon atoms. This is the vinyl chloride monomer, a gas that Monochem converts to several hundred tons of liquid each day.

A CALDRON OF GASES
Directing acetylene into a gas mixer, a worker adjusts a control valve. Hydrogen chloride, a gas supplied by an adjacent plant, also enters the vessel through another pipeline. Inside, a series of baffles creates a turbulence that mixes the two gases. The gooseneck pipe contains a safety disc designed to release the gas if pressure approaches acetylene's explosion point.

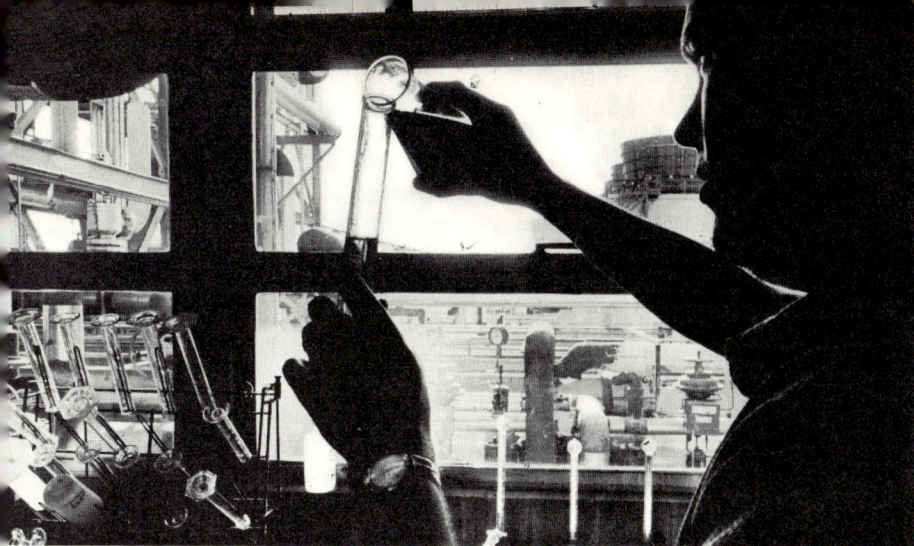

WATCHING THE WATER
Monochem's chemists inspect for impurities throughout the monomer-making process. Here water—about four million gallons a day are used—is tested for one of 22 different impurities.

AN INVISIBLE REACTION
Inside these reactors, vinyl chloride monomer is formed instantaneously when a mixture of hydrogen chloride and acetylene gases comes into contact with the catalyst, mercuric chloride.

In its last processing steps at Monochem, the gaseous monomer is purified in the towers above, and then is compressed into liquid form.

A Change from Gas to Liquid

Monochem's part in the making of Naugahyde ends when the gaseous vinyl chloride monomer is purified and compressed into liquid form for shipment. Each year the plant consumes about 25 billion cubic feet of natural gas and 90,000 tons of hydrogen chloride to make more than 250 million pounds of monomer. It also provides carbon monoxide and hydrogen to manufacture formaldehyde, widely used as a preservative and as a monomer for other synthetics.

Much of the vinyl chloride monomer, liquefied because it takes up less space and can be handled more efficiently, is retained by Uniroyal. Pumped into oversized railroad tank cars, each with a capacity of up to 40,000 gallons, the monomer is then transported 850 miles to the Painesville, Ohio, plant, where it is unloaded into spherical tanks (below) to await the next step: polymerization.

THE END OF A JOURNEY
On arrival at Uniroyal's Painesville manufacturing plant, near Cleveland, the liquefied vinyl chloride monomer is pumped from railroad tank cars into two storage tanks. The spheres, which have a combined capacity of 160,000 gallons, are painted white, reducing heat absorption that would cause the liquid monomer to expand and increase pressure inside the tank. As additional protection against excessive heat, the tanks are circled by a spiral of pipe, which is triggered like a sprinkler system to spray water on the surface automatically if a fire occurs nearby.

A Potful
of Polymers

At Painesville, Ohio, the liquid vinyl chloride monomer, made up of simple molecules, becomes transformed into the solid giant molecule of polyvinyl chloride. The conversion takes place inside polymerizers (left), 13-foot-deep vessels lined with blue cobalt glass. The polymerizers are partly filled with water to which the monomer is added; an agitator breaks the monomer into tiny droplets which disperse through the water and prevent a mass of polymer from forming.

As the catalyst—lauryl peroxide—and regulators are introduced, and the vessel is heated, the contents begin to react. When the temperature reaches about 125° F., the catalyst starts breaking the carbon double bonds in each monomer, forcing the molecules to begin linking up to one another. After a cycle that requires about eight hours, giant molecules, made up of more than 1,000 monomer units, have been created. These polymers, resembling grains of powder, are distributed from the centrally located Painesville plant as the raw material from which hosts of useful vinyl products can be manufactured.

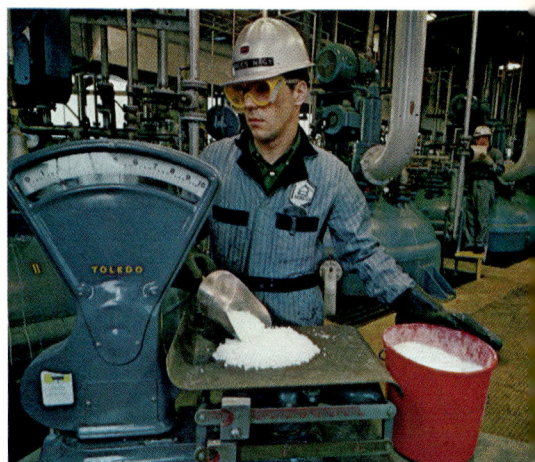

THE RECIPE FOR POLYMERS
The chemicals required to polymerize the vinyl chloride monomers are carefully measured before being added. This chemical is the catalyst, which controls the rate of polymerization.

A DEEP BLUE HOLE
Inspecting the cleaned interior of a polymerizer, a harnessed and helmeted worker descends to make sure no bits of polymer remain to contaminate the next batch. The lining of cobalt glass is used on the steel sides because it resists deterioration from the chemical reaction.

WHERE POLYMERS ARE BORN
A technician watches a reaction taking place inside a polymerizer, while a coworker (rear) bolts down the lid of a vessel that has just been cleaned. Polymerization starts as steam, circulated between the vessel's inner and outer walls, provides heat to start the reaction. Thereafter, the reaction itself gives off so much heat that cold water is circulated to control it.

SIFTING THE POLYMERS
Particles of polyvinyl chloride fall from the lip of a pan as a Painesville inspector watches for impurities. The particles, separated from the water by centrifugal action, have been dried by hot air blown through a huge rotary dryer.

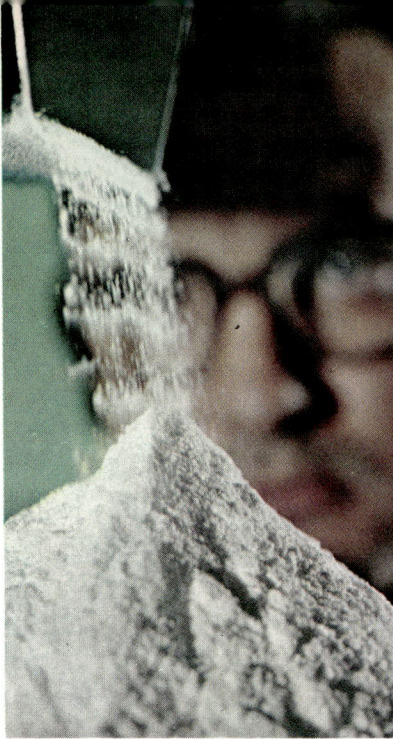

A ROUND PACKAGE
Ready for shipment, a rubber bag filled with about 9,000 pounds of polyvinyl chloride is hoisted onto a truck at Painesville while another bag is filled in the foreground. The unusual packaging is an economical way of transporting and storing large amounts of the powdery polymer, which would cluster in the corners of boxes.

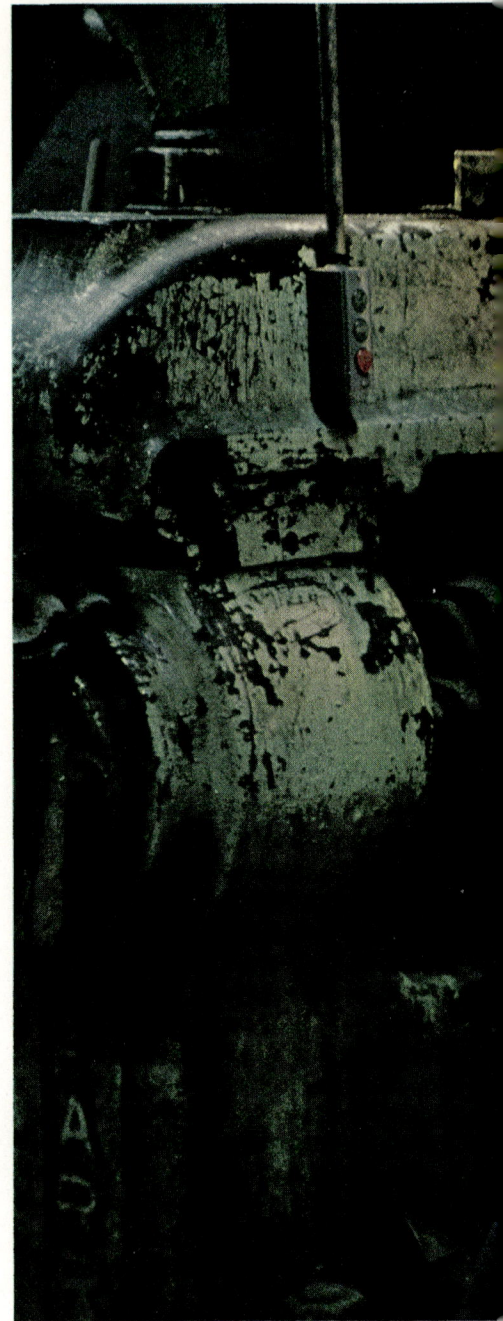

ROLLED AND REROLLED
A band of hot, sticky polymer *(above)* is stripped from rollers at Uniroyal's Mishawaka plant. The plasticized polyvinyl chloride, which becomes workable when heated to 325° F., is rolled and trimmed to the proper thickness and width before being fed into rollers which laminate it to fabric by a process called calendering.

A Powder
Becomes a Plastic

The 100 million pounds of polyvinyl chloride produced at Painesville each year are sold to manufacturers who convert it into such vinyl articles as transparent raincoats, waterproof dolls and acid-resistant plumbing. The polymer is also used to make vinyl siding for houses.

Because many products require a different form of polyvinyl chloride, 25 types are made at Painesville simply by varying the length of the polymerizing cycle, as well as the reaction temperature and the amount of agitation. The varieties differ in size, shape, weight and molecular structure. The smallest grains are 1/25,000 of an inch in diameter. Naugahyde uses a coarser grain, about 100 times larger *(left, above)*, that is transported from Painesville to Mishawaka, Indiana, in huge rubber bags *(left, below)*. There, another unit of Uniroyal, the Consumer and Industrial Products Division, begins the final processing *(above)* that each year converts about 15 million pounds of polyvinyl chloride into Naugahyde.

A Many-sided Leather

Making Naugahyde is a little like baking pie crust. First the flour—polyvinyl chloride particles—is mixed with water—in this case chemical plasticizers—to produce a plastic "dough" that makes Naugahyde either soft, for baby clothing, or tough, for upholstery fabric (right). Additional ingredients—the pigments and chemical stabilizers—color the plastic and protect it from deteriorating.

In the final production steps, the "dough" is rolled out between heated rollers, which first press the material into a thin sheet and then coat it to a cloth backing. By passing through embossing rollers, Naugahyde can emerge with a number of different finishes, including smooth, grainy or pebbled. Other processes can sandwich a layer of foam between the vinyl surface and the cloth backing, making the finished fabric as soft as chamois, or can glue fuzzy fibers to the vinyl, to make synthetic suede.

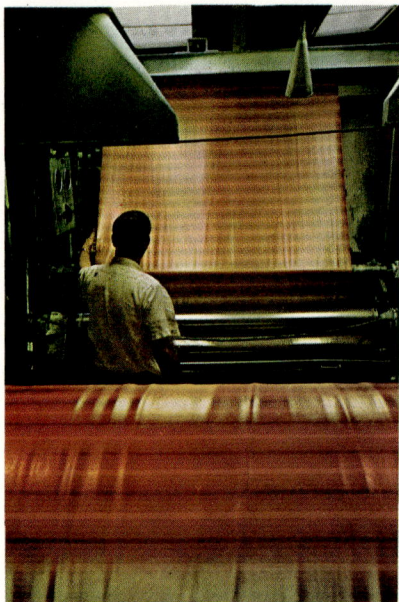

THE END OF THE LINE
Rolling off the production line at Mishawaka (right), finished Naugahyde upholstery fabric folds into a trough as a foreman checks the color accuracy. This fabric is then wound on cylinders and will be shipped to a maker of restaurant benches. In another part of the plant (above), a workman adjusts a machine printing stripes on a moving sheet of Naugahyde.

The Sign Language of the Chemist

APPENDIX

THE POLYMER CHEMIST has three ways of describing a compound. He can call it by name (e.g., methane); he can use its formula (CH_4); or he can illustrate the compound with its structural diagram (below). But names often become cumbersome (e.g., paradimethylaminoazobenzene), and while formulas tell the type and number of atoms in a compound, they do not show their physical relationship. Structural diagrams provide a detailed picture of a compound.

The key to understanding structural diagrams is a knowledge of what chemists call functional groups or linkages, units that occur regularly in organic compounds. Each group imparts its unique property to the compound it helps form and serves to classify families of such compounds. A carboxyl group, for example, always indicates an acid (diagram 5). In the series of diagrams below, all based on methane, each new functional group is shown in color.

METHANE has four hydrogen atoms linked to the four bonds of a carbon atom. In chemical reactions, one hydrogen atom (brown) can be removed, freeing a bond to join with other atoms or groups. The resulting methyl group is in black in the diagrams below.

THE BENZENE RING identifies a group of compounds known as aromatics. When linked to a methyl group, it forms toluene, used to make TNT. The ring, represented here by its symbol (brown), consists of six hydrogen atoms joined to six carbon atoms linked in a hexagon.

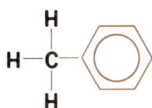

THE HYDROXYL GROUP consists of a hydrogen atom linked to an oxygen atom. When attached to a carbon atom, it forms an alcohol. This diagram—a methyl group and a hydroxyl group joined to a carbon atom and two hydrogen atoms—is of ethyl alcohol.

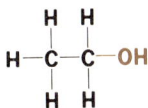

THE CARBONYL GROUP consists of a carbon atom double-bonded to an oxygen atom, and appears in such compounds as formaldehyde and nail polish remover. When linked to a methyl group and a hydrogen atom, the result is ethanal, used in drug production.

THE CARBOXYL GROUP, present in all organic acids, consists of a carbon atom double-linked to an oxygen atom and single-linked to a hydroxyl group. When it is joined to a methyl group, it forms acetic acid, a major constituent of vinegar.

THE ESTER LINKAGE is formed of a carbon atom double-bonded to one oxygen atom and single-bonded to another. In this diagram the carbon atom joins with a methyl group while the free bond of the oxygen atom holds a benzene ring. The ring, in turn, links with a carboxyl group. The compound formed is acetylsalicylic acid, or aspirin. When a series of related monomers is joined through the ester linkage, the result is a polyester molecule, from which plastics such as Mylar are made.

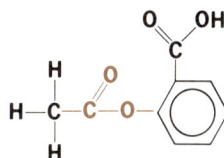

THE ETHER LINKAGE, a functional "group" that consists of a single oxygen atom, is always bonded on either side to a carbon atom. To create anisole (below), which is used in making perfume and is a relative of the anesthetic ether, the linkage is joined to a methyl group on one side, to a benzene ring on the other.

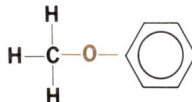

URETHANE AND UREA LINKAGES combine with several other groups to form a constituent of certain synthetic rubbers. Attached to the carbon atom are two methyl groups. To the left is a benzene ring, and at far left is a urethane group, one of the several organic groups that contain nitrogen. To the right of the methyl groups is another benzene ring followed by a urea group (shaded). Attached to the urea group is a third benzene ring, which, in turn, is attached to a second urethane group, a mirror image of the first urethane group but chemically exactly the same.

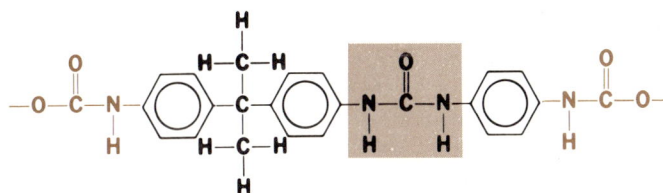

193

A Polymer Primer

THE WORLD OF GIANT MOLECULES encompasses three branches of science: organic chemistry, physical chemistry and polymer chemistry. The pioneers in the field, the organic chemists, contributed their vocabulary, adding new words as new substances were created. When the physical chemists entered, in their role as investigators of the physical properties of atoms and molecules, the list of technical terms grew larger. Finally the polymer chemists, specialists in giant molecules, who borrowed from the organic and physical chemists, created additional terms. The result has been a special language numbering in the thousands of words, few of them familiar to the layman. The glossary that follows defines some of the most important of these words.

ACRYLIC. The name applied to the group of synthetic materials whose basic units, or monomers, include acrylic acid ($C_3H_4O_2$) or its derivatives. Acrylic materials include synthetic fibers such as Acrilan and Orlon, and plastics such as Lucite and Plexiglas.

ADDITION POLYMERIZATION. One method of combining molecular building blocks, or monomers, into long chain molecules, or polymers. This type of polymerization is generally accomplished by subjecting the monomers to heat and pressure. All the original components appear in the resulting substance—there are no castoffs or by-products (see CONDENSATION POLYMERIZATION).

BENZENE RING. A molecular structure composed of six atoms of hydrogen and six atoms of carbon. Each hydrogen atom is linked to a carbon atom, and the carbon atoms are arranged in a hexagon. The bonds between the carbon atoms are unusual in that they are neither single nor double; they are hybrid links, a cross between single and double—meaning that some of their electrons, rather than being shared by just two atoms, are shared by all the carbon atoms in the ring. Benzene rings are found in the compounds that form such synthetic materials as epoxy glues (see COVALENT BOND).

BLOCK POLYMER. A giant molecule composed of sets of chemically different monomers. In such a structure monomers are first linked to form short chains, called blocks. The different blocks are then joined in a regular pattern, i.e., AAAA . . . BBBB . . . AAAA . . . BBBB . . . etc. (see POLYMER and COPOLYMER).

BRANCHED CHAIN MOLECULE. A polymer whose molecules are linked linearly—i.e., in a straight chain—but with additional monomers adhering at angles to the basic chain, forming short branches. Materials made of branched chain molecules are generally less dense, more flexible and less resistant to heat than those made of unbranched chain molecules (see UNBRANCHED CHAIN MOLECULE).

CATALYST. A substance whose properties allow it to accelerate certain chemical reactions, but which does not appear in the product of such reactions. Although a catalyst becomes temporarily involved in a chemical reaction, it emerges unchanged (see INITIATOR).

CONDENSATION POLYMERIZATION. A method of joining monomers to form giant molecules, in which all the elements present do not appear in the final substance. Water or other simple compounds are the usual by-products of the process (see ADDITION POLYMERIZATION).

COPOLYMER. A giant molecule composed of two or more chemically different monomers (see POLYMER and BLOCK POLYMER).

COVALENT BOND. The more common of the two major forces by which atoms are joined to form a chemical compound. It occurs when the atoms of a compound share pairs of electrons. When two atoms share one pair of electrons, the bond is called a single bond; sharing two pairs of electrons results in a double bond. The maximum number possible is the triple bond (see IONIC BOND).

CROSS-LINKED MOLECULE. A molecular structure in which adjacent polymer chains are joined by short molecular chains. Cross-linkage—sometimes called "network structure"—increases the strength and heat resistance of a substance, but at the same time makes the substance more rigid, and therefore more difficult to work with. It creates, in effect, a single immense molecule. Bakelite is a cross-linked polymer.

CRYSTALLINITY. The property of a substance which derives from the three-dimensional arrangement of its atoms, ions or molecules in a regular pattern. In diamonds, for example, which are composed only of carbon, the atoms are usually arranged in interconnected octagons.

CYCLIC MOLECULE. A molecule which appears in organic compounds, and whose atoms are arranged in a ring, or polygon. One of the most frequently occurring cyclic structures in synthetic polymers is the six-sided benzene ring (C_6H_6) (see BENZENE RING).

ELASTOMER. A substance which can be stretched and which returns to its original shape. One type of elastomer is synthetic rubber.

FIBRID. A synthetic fibrous structure used as a substitute for wood pulp (cellulose) in the production of a synthetic paper, and in certain nonwoven synthetic fabrics.

FLUOROCARBON. A compound composed mainly of carbon and fluorine. Fluorocarbons are analogous to hydrocarbons, but with atoms of fluorine substituted for hydrogen atoms. Such compounds are far stronger and more highly heat resistant than hydrocarbons. Teflon, the coating applied to nonstick cooking ware, is a fluorocarbon.

HYDROGEN BOND. A weak bond between molecules that links a hydrogen atom in one with two atoms, generally of oxygen or nitrogen, in the other. The attraction comes from the opposite electrical charges of the atoms involved. Such a bond exists between molecules of nylon, and also between molecules of water.

INITIATOR. A substance whose presence is necessary to the process of building certain giant molecules. An initiator helps bring about such reactions and—unlike a catalyst—is changed by them and may appear in the final product (see CATALYST).

IONIC BOND. The attraction that results when one atom gives up electrons to another atom or group of atoms. Atoms that thus lose or gain electrons are called ions. While atoms are electrically neutral, ions have either a positive or a negative charge. It is the mutual attraction of oppositely charged ions that causes the formation of an ionic bond. Sodium and chlorine, the elements of table salt, are linked by an ionic bond (see COVALENT BOND).

ISOMERS. Molecules belonging to families that contain exactly the same type and number of atoms, but with the atoms differently arranged in each case. For example, butane and isobutane share the same formula (C_4H_{10}), but their structural diagrams would show differences in the placement of the atoms. One type of paraffin molecule ($C_{40}H_{82}$) has more than 62 trillion possible isomers.

MOLECULAR WEIGHT. The sum of the weights of the atoms in a given molecule. Thus since hydrogen (H) has an atomic weight of one, and chlorine (Cl) a weight of 35.5, a molecule of hydrochloric acid (HCl) has a molecular weight of 36.5.

PLASTIC. A synthetic material composed of giant molecules which contain atoms of carbon, as well as atoms of other elements such as hydrogen, oxygen, nitrogen and silicon. A diversity of products, from epoxy glue to polyethylene bags, are plastics.

POLYAMIDE. The technical name for the various kinds of nylon, whose molecules are characterized by recurring groups of carbon, oxygen, nitrogen and hydrogen (—CONH—), in an arrangement called an amide.

POLYCARBONATE. A class of plastics in which the physical arrangement of the carbon and oxygen atoms in a special grouping (—OCOO—) is such that the resulting product combines the qualities of transparency, temperature resistance, electrical nonconductance, rigidity and great strength.

POLYESTER. A large group of synthetic compounds produced by combining certain types of acids and alcohols. Some polyesters are drawn into fibers, such as Dacron and Kodel; others are used to make reinforced plastic skis.

POLYMER. A giant molecule formed of up to hundreds of thousands of smaller molecules in a regular pattern.

REINFORCED PLASTICS. Synthetic materials mixed with or laminated to other materials for increased strength. Certain polyester plastics, for example, are reinforced by glass fiber for use in boat hulls or automobile bodies; the result of this is often lighter than aluminum, yet stronger than steel.

RESIN. A substance which occurs in nature as the amber-colored, sticky secretion of certain insects, plants and trees, and in polymer chemistry as the intermediate, solid or semisolid stage of all plastics and synthetic fibers.

SATURATED MOLECULE. A molecule in which each bond of every atom in the molecule's backbone holds another atom.

SILICONES. A family of synthetic materials based on the elements silicon and oxygen. Because of their molecular structure—a backbone composed of alternating atoms of silicon and oxygen, usually with associated carbon atoms—many silicones are characterized by an ability to withstand extremely high temperatures that makes them useful, among other things, as lubricants in jet aircraft engines.

UNBRANCHED CHAIN MOLECULE. A giant molecule whose components line up in a row; also called a straight chain molecule (see BRANCHED CHAIN MOLECULE).

VALENCE. The characteristic of an atom that determines how many other atoms of a given element it may combine with. Valence is established by the number of electrons an atom loses, gains or shares when combining with other atoms to form compounds. Thus an atom of carbon, with a valence of four, combines with four hydrogen atoms, each with a valence of one, to form methane (CH_4).

VINYL PLASTIC. Any plastic whose molecules are formed from vinyl units; that is, units, or monomers, in which the last two carbon atoms at either end are connected by a double bond (C=C) instead of merely a single bond (C—C). Because one of these bonds may be broken to form a new bond with another unit, vinyl molecules are almost unexcelled as synthetic building blocks.

VULCANIZATION. The process of chemically combining rubber with sulfur to eliminate the rubber's tendency to become tacky when warm and brittle when cool.

FURTHER READING

General

Choppin, Gregory R., and Bernard Jaffe, *Chemistry.* Silver Burdett, 1965.

*Gamow, George, *Matter, Earth and Sky* (2nd edition). Prentice-Hall, 1965.

Lapp, Ralph E., and Editors of LIFE, *Matter.* Time Inc., 1965.

*Lessing, Lawrence, *Understanding Chemistry.* John Wiley & Sons, 1959.

Pauling, Linus, and R. Hayward, *The Architecture of Molecules.* W. H. Freeman, 1964.

†Saunders, B. C., and R.E.D. Clark, *Atoms and Molecules Simply Explained.* Dover Publications, 1964.

Sienko, Michell, and Robert A. Plane, *Chemistry* (2nd edition). McGraw-Hill, 1961.

History of Chemistry

Dorian, Max, *The du Ponts.* Little, Brown, 1962.

†Editors of *Scientific American, Lives in Science.* Simon and Schuster, 1957.

Findlay, Alexander, *A Hundred Years of Chemistry* (3rd revised edition). Gerald Duckworth, 1965.

Ihde, Aaron J., *The Development of Modern Chemistry.* Harper & Row, 1964.

†Jaffe, Bernard, *Crucibles, the Story of Chemistry.* Fawcett Publications, 1960.

*Partington, J. R., *A Short History of Chemistry.* Harper & Row, 1960.

Taylor, F. Sherwood, *A History of Industrial Chemistry.* Abelard-Schuman Ltd., 1957.

Organic Chemistry

*Asimov, Isaac, *The World of Carbon.* Abelard-Schuman Ltd., 1958.

Fieser, Louis, and Mary, *Introduction to Organic Chemistry.* D. C. Heath, 1957.

†Gibbs, F. W., *Organic Chemistry Today.* Penguin Books, 1961.

Giant Molecules

Jaffe, Bernard, *Chemistry Creates a New World.* Thomas Y. Crowell, 1957.

Moncrieff, R. W., *Man-made Fibres* (4th edition). John Wiley & Sons, 1963.

†O'Driscoll, Kenneth F., *The Nature and Chemistry of High Polymers.* Reinhold, 1964.

Scientific American, "Giant Molecules" (entire issue). September 1957.

Simonds, Herbert R., and James M. Church, *A Concise Guide to Plastics* (2nd edition). Reinhold, 1963.

Stille, J. K., *Introduction to Polymer Chemistry.* John Wiley & Sons, 1962.

Wolf, Ralph F., "Seventy-Five Year Stretch." *Rubber World,* October 1964.

---

*Also available in paperback edition.

†Only available in paperback edition.

ACKNOWLEDGMENTS

The editors of this book are especially indebted to Murray Goodman, Professor of Chemistry, Polytechnic Institute of Brooklyn, and to the following persons and institutions: Roy Avery, Director of Public Relations, American Chemical Society, New York; Judith Bregman, Associate Professor of Physics, Polytechnic Institute of Brooklyn; William J. Connelly, Manager of Advertising, and his associates, Union Carbide Corp., Plastics Division, New York; William Cox, Public Relations Dept., Celanese Corp. of America, New York; W. Crouch, Director of Synthetic Rubber and Plastics Research, Phillips Petroleum Co., Bartlesville, Okla.; Derham-Miles Associates, Teflon Ice Skating Rink, Philadelphia; the following people from E. I. du Pont de Nemours & Co., Inc.—Barbara Gibbon, Product Information & Textile Fibers, H. V. Hardy, Public Relations Dept., Jack D. Hunter, Editor, *Du Pont Magazine,* John F. Loutz, Research Associate, Plastics Dept., Domenico Mortellito, Design Adviser, Alexander L. Murphy, Public Relations, Visual Aids Service Dept., Gordon D. Patterson Jr., Analytical Supervisor, Film Research Laboratory, and John O. Punderson, Senior Research Chemist, Plastics Dept.; Roger Durant, Public Information Dept., Dickinson College, Carlisle, Pa.; Enjay Chemical Co., Division of Humble Oil & Refining Co., New York; Lawrence Faeth, Photographic Supervisor, Olin Mathieson Chemical Corp.; Martin Falxa, Polytechnic Institute of Brooklyn; John Freilich, University of Connecticut; Michael Fried, Research Assistant, Polytechnic Institute of Brooklyn; William Fuss, owner, William Fuss (floral decorators), New York; the following people from General Electric Co.—Stewart Brewer, Manager, Resin Product Development, W. E. Harris, Advertising & Sales Promotion, Lawrence Mahar, Advertising & Public Relations, Vincent Manti, Research & Development Center, Robert J. Prochaska, Manager, Polycarbonate Research & Development, Walter Robb, Manager, Chemical Processes Branch, and Peter Van Avery, Public Information; Barrett K. Green, Manager of Chemical Research, National Cash Register Co., Dayton; Harry P. Gregor, Professor of Physical Chemistry, Polytechnic Institute of Brooklyn; Richard J. Griffith, Fred Rosen Assoc., Public Relations Counsel, National Cash Register Co., New York; P. F. Gunberg, Manager, Technical Service for Sales, Texas-U.S. Chemical Co.; Robert Hedges, Allied Chemical Corp.; Everett Holt, Tacoma Park, Md.; M. W. Kellog Co., New York; Richard M. Klein, New York Botanical Gardens, Bronx Park, New York; Bert S. Kleinsinger, Bronx High School of Science, New York; Sylvan Kling, Public Relations Account Manager, Arndt, Preston, Chapin, Lamb & Keen, Inc., Philadelphia; Frederick C. Kreiling, Associate Professor of History, Polytechnic Institute of Brooklyn; Fred Lentz, Manager, Rayon Quality Control, American Enka Corp., Asheville; Harvey C. McCormick Jr., Public Relations, The Borden Chemical Co., New York; William B. deMeza, Public Relations Dept., The Goodyear Tire & Rubber Co., Akron; the following people from Monochem, Inc., Geismar, La.—William J. Boyne, R. H. Hamilton and Jack R. Little; James E. Murphy, Public Relations Manager, Owens-Corning Fiberglas Corp., Toledo; Natural Rubber Bureau, Washington; James B. Ogdin, Sales Coordinator, Eastman Chemical Products, Inc., Kingsport, Tenn.; Joseph P. Orr, Plastic & Packaging Publicity, Monsanto Co., St. Louis; William M. Rhodes, Coordinator of Photography, Sun Oil Co., Philadelphia; Thomas E. Roney, Technical Director, Marbon Chemical Division, Borg-Warner Corp., Washington, W. Va.; J. H. Saunders, Director of Research, Mobay Chemical Co., Pittsburgh; Lowell Schliecher, Fundamental Research, National Cash Register Co., Dayton; Scientific Design Co., New York; Phyllis Stevens, New York; Elma St. John, Fashion Coordinator, Tovar Tresses, New York; the following people from U.S. Rubber Co.—Edward Loeffler and P. J. Sweeney, Mishawaka, Ind.; Robert Bailey Jr., Tom Doherty, Charles D. McCleary, Research Director, J. M. Poynter, Duncan Ross, Public Relations, and Arnold Werner, Naugatuck Chemical Division; Richard V. Vratanina, South Bend, Ind.; and Elsie Watson and associates, Public Relations Dept., Firestone Tire & Rubber Co., Akron.

INDEX

Numerals in italics indicate a photograph or painting of the subject mentioned.

W

Water, 10, 52; Aristotle's concept of, 31; conservation, 175; first synthesis of, 32; treatment techniques, *118-119*, 169, 174-175
Water-and-oil repellence, of fluorocarbon polymers, 145, 149

Waxes, 23, 53, *61*, 79
Wickham, Henry, 125
Wilhelm II, King of Prussia (German Kaiser), 52
Williams, Greville, 125
Windowpanes, unbreakable, *19*, 156, 173

Wöhler, Friedrich, 34, 35, 46, *47*
Wood, 12, 33, 34, 56, 132; cellulose in hemlock chip, *85;* cellulose content of, 74, 84; cellulose production from, 76, *83-85*, 86; pulp, 82, 84, 92, 153
Wool, 12, 56; Orlon compared to, *119*

X

X-ray diffraction, 100-*101*

Z

Zepel, 149

PICTURE CREDITS

The sources for the illustrations which appear in this book are shown below. Credits for the pictures from left to right are separated by commas, from top to bottom by dashes.

Cover—John Zimmerman.

CHAPTER 1: 8—Leviton-Atlanta for FORTUNE. 10—The Granger Collection. 13—Drawings by George V. Kelvin. 17—Jerome Ducrot. 18, 19—Ralph Crane except top right E. I. du Pont de Nemours & Company, Inc. 20—Bruce Roberts from Rapho-Guillumette. 21—Seymour Mednick. 22, 23—Richard Meek for SPORTS ILLUSTRATED. 24, 25—The Goodyear Tire and Rubber Company. 26—Arnold Newman. 27—Arnold Newman—David Holdridge courtesy Phyllis Stevens. 28, 29—Gordon Tenney.

CHAPTER 2: 30—Derek Bayes courtesy Trustees of the British Museum. 32—Courtesy The Burndy Library, Inc., Norwalk, Connecticut. 35—Deutsches Museum, München. 36—Eddy Van der Veen courtesy Bibliothèque Nationale. 37—Historiches Bildarchiv-Bad Berneck. 39—Donald Miller. 40, 41—A. G. Ingram Ltd. courtesy University of Edinburgh—Derek Bayes courtesy The Royal Scottish Museum, Edinburgh, Donald Miller. 42—Ben Rose. 43—Henry Groskinsky, Heinz Zinram courtesy Dr. Williams Library, London. 44, 45—Eddy Van der Veen courtesy Conservatoire des Arts et Métiers, Donald Miller—Eddy Van der Veen. 46—Walter Sanders courtesy Anorganic Chemical Institute, Göttingen. 47—Donald Miller, Friedrich Rauch, München—Ben Rose. 48—Ben Rose. 49—Heinz Zinram courtesy Trustees of the National Portrait Gallery, London.

CHAPTER 3: 50—General Electric Company. 52, 54, 56—Drawings by George V. Kelvin. 59 through 62—Ken Kay. 63—Dow Chemical Corporation. 64 through 71—Ken Kay.

CHAPTER 4: 72—Ken Kay courtesy Union Carbide Plastics Division. 74, 75—Drawings by George V. Kelvin. 77—Bibliothèque du Museum d'Histoire Naturelle. 78—Drawings by George V. Kelvin. 79—Drawings by Leslie Martin. 80—Courtesy Union Carbide Plastics Division—drawings by George V. Kelvin. 83—Charles Mikolaycak. 84, 85—International Paper Company, United States Department of Agriculture, Forest Products Laboratory, Madison, Wisconsin—Professor R. D. Preston, The Astbury Department of Biophysics, University of Leeds. 86—Culver Pictures, Celanese Corporation of America—Bill Snead for Topeka

Capital Journal. 88—Drawings by Nicholas Fasciano. 89 through 97—Bruce Roberts from Rapho-Guillumette.

CHAPTER 5: 98—Max B. Miller. 101—Courtesy Polaroid Corporation, Cambridge. 102, 103—Drawings by Joseph del Gaudio. 106, 107—Drawings by George V. Kelvin. 109—E. I. du Pont de Nemours & Company, Inc. 110—The Eleutherian Mills Historical Library. 111—The Hagley Museum—E. I. du Pont de Nemours & Company, Inc. 112—Culver Pictures—E. I. du Pont de Nemours & Company, Inc. 113, 114, 115—E. I. du Pont de Nemours & Company, Inc. 116, 117—E. I. du Pont de Nemours & Company, Inc. except top right Gabriel Benzur. 118, 119—E. I. du Pont de Nemours & Company, Inc. 120—Jerry Cooke courtesy E. I. du Pont de Nemours & Company, Inc. 121—John Launois from Black Star.

CHAPTER 6: 122—The Warshaw Collection of Business Americana. 124—Drawings by Leslie Martin. 126—Drawings by George V. Kelvin. 127—The New York Public Library. 128—The Bettmann Archive. 129—Drawings by George V. Kelvin. 130, 131—Drawings by Donald and Ann Crews. 133—Hamilton Wright. 134 through 143—Drawings by Otto van Eersel.

CHAPTER 7: 144—Ted Russell. 146, 147, 149—Drawings by George V. Kelvin. 151—Professor Stig Claesson, University of Uppsala, Sweden. 155, 156, 157—John Zimmerman. 158—Top John Zimmerman; bottom Eastman Chemical Products Inc. 159—John Zimmerman. 160—Vernon Pope Company, Inc. 161—Donald Miller. 162, 163—E. I. du Pont de Nemours & Company, Inc., Donald Miller. 164—John Zimmerman. 165—Drawing by Nicholas Fasciano, John Zimmerman. 166, 167—E. I du Pont de Nemours & Company, Inc., John Zimmerman.

CHAPTER 8: 168—Ivan Masser from Black Star. 173—Courtesy The National Cash Register Company, Dayton, Ohio—courtesy World Publishing Company, Cleveland, Bible # 714. 175—Drawing by Nicholas Fasciano. 177—Ralph Morse. 179 through 183—Gordon Tenney. 184—Anthony Wolff. 185 through 191—Gordon Tenney. Back cover—Drawing by Nicholas Fasciano.

A STONEHENGE BOOK

XXXXX Printed in U.S.A.